Environmental Footprints and Eco-design of Products and Processes

Series editor

Subramanian Senthilkannan Muthu, SGS Hong Kong Limited,
Hong Kong, Hong Kong SAR

More information about this series at http://www.springer.com/series/13340

Subramanian Senthilkannan Muthu
Editor

Environmental Footprints of Packaging

Springer

Editor
Subramanian Senthilkannan Muthu
Environmental Services Manager-Asia
SGS Hong Kong Limited
Hong Kong
Hong Kong SAR

ISSN 2345-7651 ISSN 2345-766X (electronic)
Environmental Footprints and Eco-design of Products and Processes
ISBN 978-981-10-1236-5 ISBN 978-981-287-913-4 (eBook)
DOI 10.1007/978-981-287-913-4

Springer Singapore Heidelberg New York Dordrecht London

Softcover re-print of the Hardcover 1st edition 2016

Printed on acid-free paper

Springer Science+Business Media Singapore Pte Ltd. is part of Springer Science+Business Media
(www.springer.com)

Preface

Packaging is one of the essential elements of today's life, and it plays a major role in our daily lives. Packaging is inevitable to different communities of people: manufacturers, shopkeepers, sellers, consumers and so on. Several kinds of packaging are made out of a wide array of materials we are surrounded with every day. Packaging serves an essential function of protecting goods from damage, apart from the other secondary functions, and it is used by every industrial segment. The environmental impacts of any product produced on Earth deserve significant attention these days, and this attention is very high for packaging because of its voluminous applications. Due to this, one can imagine the quantity of production of packaging materials and the associated environmental impacts. Not only the production of packaging, but also its disposal, creates impacts to the environment. Many environmental elements—such as the biodegradation potential of packaging materials, the uncountable proportion of consumption and disposal, the short shelf-life of packaging materials, and limited landfill space, etc.—are associated with this issue.

The dissemination of information and the knowledge of quantification of environmental footprints of different packaging materials and packaging systems are of great benefit to concerned consumers as well as researchers in the scientific community, and this book is an attempt toward the same. This book deals with the environmental footprints of packaging in seven informative chapters. All seven chapters deal with various important elements associated with the environmental implications of packaging: (1) the life-cycle assessment of packaging systems; (2) the sustainable design of packaging materials; (3) organization–life cycle assessment (OLCA); methodological issues and case studies in the beverage-packaging sector; (4) the potential of fibrous and nonfibrous materials in biodegradable packaging; (5) the environmental impacts of packaging materials; (6) the bioprocessing of metals from packaging wastes; and (7) the environmental implications of reuse and recycling of packaging. I am sure that the readers of the book will receive much useful information pertaining to the environmental footprints of packaging. I take this opportunity to thank all of the authors who contributed the chapters in this book for their time and priceless efforts.

Contents

Life-Cycle Assessment of Food-Packaging Systems

Giuseppe Vignali

Abstract Food packaging plays a fundamental role in today's society because it protects food from external sources of contamination and preserves food properties during the entire assigned shelf life. Due to this fundamental role, its use is increasingly widespread including emerging or underdeveloped countries. The global amount of packaging materials manufactured and disposed of every day has led many researchers to deal with the issue of their environmental impact. Several studies have been performed starting from 1990 to the present that have been aimed at demonstrating the best type of use of and end of life for each type of food-packaging material. In recent years, some studies have also demonstrated how the extension of the food shelf life by means of improved packaging could decrease the environmental impact of an entire packaged food based mainly on the reduction of the associated food waste. Based on these premises, this chapter aims at reviewing the main articles in the field of environmental assessment of food packaging by means of a life-cycle assessment approach and showing how, during the last two decades, this issue has received increasing attention. The review was performed by analysing 172 scientific papers collected from the Scopus database using specific keywords and refining the results based on a detailed analysis of the content of each article. The results show how interest in this topic has grown consistently during the last 25 years and indicates several research lines available for further studies in this field.

Keywords Food packaging · LCA · Environmental impact · Food waste · Review · Packaging systems

G. Vignali (✉)
Department of Industrial Engineering, University of Parma, Viale Delle Scienze 181/A, 43124 Parma, Italy
e-mail: giuseppe.vignali@unipr.it

S.S. Muthu (ed.), *Environmental Footprints of Packaging*,
Environmental Footprints and Eco-design of Products and Processes,
DOI 10.1007/978-981-287-913-4_1

1 Introduction

In today's society, packaging has a key role in sustainability: It is no longer possible for people involved in the design, development, production, or use of packaging to ignore the environmental consequences of their work (Almeida et al. 2010). Packaging activities and materials, rather than just only the final result visible to the consumer, should be considered in each phase of a product supply chain.

The packaging sector generates approximately 2 % of the gross national product in developed countries, and approximately half of this packaging concerns the food sector (Robertson 2012). On average, in 2012, European citizens generated globally 80 million tons of packaging waste, 38 % of which was paper, 21 % plastic, 20 % glass, 15 % wood, and 6 % metals (Eurostat 2014). Overall, the packaging material life cycle generates significant environmental impacts; indeed, its production and application to food products requires the use of natural resources and energy and causes relevant emissions. Moreover, packaging wastes generate increasing disposal wastes, with the first largest fraction being municipal waste, that have exceeded the organic fraction in many countries (Edjabou et al. 2015). In developed counties, modern end-of-product life management systems can strongly reduce the environmental impacts of packaging; however, in many underdeveloped or developing countries, waste-management systems are fairly rudimentary (Denafas et al. 2014; Sealey and Smith 2014) and create significant environmental problems.

To measure the "green" characteristics of a product or a system, it is essential to assess its environmental impacts and resources utilization using quantitative and objective methodologies that consider its entire life cycle. In this regard, life-cycle assessment (LCA), as regulated by the ISO 14040 (2006) International Series of Standards, is a useful methodology to assess the environmental impact of a product throughout its lifetime. This methodology allows quantifying the level of greenhouse gas emissions, the amount of energy consumed, and the level of hazardous substances emitted throughout a product's life cycle. It also allows, by means of a selection of specific mid- or end-point indicators, identifying the most relevant impact among those evaluated (e.g., ozone depletion, CO_2 emissions, etc.) as well as identifying the processes that generate the greatest environmental impact; the final purpose is to propose guidelines for the improvement of the current situation.

In recent years, various LCA studies related to packaging have been performed with many of them focusing on the food sector. In this field, the impact of packaging-materials production, packaging application, and packaging use and disposal have been thoroughly analysed among several food supply chains. During the latter period, evaluation of the advantages resulting from decreased food losses as a consequence of improved food packaging use has emerged (Williams and Wikström 2011; Grönman et al. 2013; Wikström et al. 2014). Depending on the type of food, the relative impact of the packaging may change reaching, e.g., a high percentage for beverages or vegetables foods (Roy et al. 2009; Manfredi and

Vignali 2014). Aware of these issues, some researchers attempted to define frameworks or guidelines to develop sustainable packaging focusing not only on the impact of the packaging materials but also on the preservation of its performance and the impact of packaging technologies (Sonneveld 2000; Grönman et al. 2013; Toniolo et al. 2013; Wever and Vogtländer 2013; Manfredi and Vignali 2015).

Based on this chapter aims to underscore the importance of the environmental impact of food-packaging systems by means of an in-depth analysis of the existing literature on this issue. A review of available data about the environmental impact of food-packaging systems was performed by dividing the reviewed works among the most frequent topics addressed in the literature. The results show how attention paid to food-packaging systems by evaluating their environmental impact has increased in the last 25 years, thus reaching a wide level of diffusion.

The remainder of the chapter proposes a description of the LCA principles applied in the context of packaging systems, describes the adopted review methodology, and presents the main results of the review as well as discussion. Finally, a section on conclusions and future research recaps the main chapter findings and, based on them, proposes new research in the field of the environmental assessment of food-packaging systems.

2 LCA of Packaging Systems

LCA is a technique to assess and quantify the environmental impacts associated with a product, process, or activity. The entire life cycle of a product, from raw-material extraction to disposal, is considered.

This method is composed of the following steps:

1. Goal and scope definition
2. Inventory analysis
3. Impact assessment
4. Interpretation and conclusion

As far as the issue of this article is concerned, the following subparagraphs are aimed at describing a generic framework as well as some particularities about applying an LCA to packaging systems.

2.1 Goal and Scope Definition

The first activities that should be performed in an LCA are (1) identifying the functional unit (i.e., the object of LCA analysis); (2) defining the study motivation (i.e., the targeted evaluation); and (3) determining the final recipients of the results (i.e., who makes use of the analysis).

2.1.1 Functional Unit

All of the problem data must be referred (for normalization) to a functional unit, e.g., in the field of food-packaging systems, a functional unit could be 1 kg of packaged food product or 1 kg of packaging materials. The choice of this reference unit is fundamental in determining how the materials flows inside the system boundaries.

2.1.2 System Boundaries

LCA analysis requires identifying the processes and activities comprising the system under examination. For each operation, the input and output should also be identified. The definition of system boundaries implicitly describes what is included and what is excluded from the analysis. Regarding the LCA of a packaged food product, the system boundaries usually include the cultivation, manufacturing and processing of the food product, as well as the packaging and distribution phases, to understand the relevant environmental impacts of the different phases. If the reference unit is related to a specific quantity of packaging materials, the system boundaries could be defined not considering food production and processing and sometimes not considering the distribution phase but only considering the packaging materials supply chain. Usually the equipment-manufacturing phase is not included in the system boundaries due to the limited impact caused by the long life time of packaging equipment. Figure 1 presents an example of system boundaries considering all possible approaches adopted in an LCA of food-packaging systems

2.2 Life-Cycle Inventory (LCI) Data Analysis

LCI analysis involves creating an inventory of flows from and to nature considering the previously selected system boundaries. Inventory flows include inputs of water, energy, and raw materials and releases to air, land, and water. The inputs and outputs are collected for all activities within the system boundaries. Packaging systems must be accurately evaluated and shown in the analysis, thus allowing comparison between different materials or processes.

There can be two types of LCI data:

- Primary data, i.e., plant-specific data
- Secondary data, i.e., average data contained in databases

The use of primary data is fundamental to assess the impact of a specific packaging system, and for this reason the use of secondary data is limited to only some minor impact or to impacts that do not vary from case to case.

Fig. 1 Possible systems boundaries regarding LCA of food-packaging systems

2.3 Life-Cycle Impact Assessment (LCIA)

LCIA is aimed at evaluating the significance of potential environmental impacts based on the LCI flow results. A classic LCIA consists of the following mandatory elements:

2.3.1 Selection of Impact Categories and Characterization Models

Some relevant impact categories internationally accepted for the evaluation of the environmental impact of food-packaging systems include the following:

- global warming potential (GWP)
- natural resource depletion
- stratospheric ozone depletion

- acidification
- photochemical ozone creation
- eutrophication
- human toxicity
- aquatic toxicity

In an LCA for food-packaging systems, various characterization models can be used to calculate the results of impact assessment.

Regarding methods to assess impacts, some of the most used methods for assessing food-packaging systems include the following:

- CML 2001 (Guinée et al. 2001)
- ReCiPe 2008 (Goedkoop et al. 2009)
- ILCD 2011 (European Commission's Joint Research Centre 2010)
- Impact 2002+ (Jolliet et al. 2003)

Classification stage is another mandatory element. In this phase, the inventory parameters are sorted and assigned to specific impact categories. Usually this phase is performed by specific commercial software frequently used for LCA of food-packaging system (e.g., SimaPro or GaBi).

Impact measurement is where LCI flows are characterized into common equivalence units, which are then summed to provide an overall impact category.

2.4 Interpretation of Results

This step consists of presenting the results of LCA analysis with the aim to:

1. identify the most relevant impact (e.g., ozone depletion, CO_2 emissions, etc.);
2. identify processes that generate the greatest environmental impact (e.g., manufacturing of packaging materials, food-packaging operations, packaging transport);
3. Propose guidelines for improvement.

3 Methodological Approach to Literature Review

The methodology adopted to identify the studies analysed in this book is systematic literature review (Transfield et al. 2003). A systematic literature review is commonly adopted to identify key scientific contributions to a field or question, and it is grounded on a rigorous, replicable, scientific, and transparent process (Cook et al. 1997).

A systematic review requires two subsequent steps (Alderson et al. 2004). First, the inclusion criteria for the selection of the studies to review should to be identified. In our case, we decided to include in the review only studies that met the following criteria:

- We selected only papers concerning LCA of packaging systems that were applied exclusively in the food sector. Therefore, articles describing the environmental assessment of packaging not applied directly to food (e.g., secondary and transport packaging) were not retained;
- We selected only papers that were published in peer-reviewed international journals. Other publication forms (e.g., books, conference proceedings, newspapers articles, unpublished works, doctoral dissertations, etc.) were not considered in order to maintain a high scientific level and also to have access to the full articles;
- We selected only papers that were written in English.

No specific criteria were defined for the publication time span.

The second step of a systematic review is the strategy of locating and selecting the studies. In our review, we performed a computerized search of the Scopus database (www.scopus.com) by entering different pairs of keywords and taking into account the different terminology used by authors when referring to the environmental assessment of food packaging. In particular, two searches were performed (1) by coupling "food" with "packaging" with either "environmental assessment" or "life-cycle assessment" as keywords (search 1); and (2) by searching for these keywords in title, abstract, or article keywords in the Scopus database (search 2). These queries lead, respectively, to 250 and 115 studies published up to June 2015 including, e.g., books, conference proceedings, newspapers articles, unpublished works, doctoral dissertations, etc. According to the above-mentioned criteria, 184 articles of the 250 obtained from search 1 and 94 articles of the 115 obtained from search 2 were considered, gathered, and checked to avoid duplication. As a result, 95 articles were removed from the original set of 315. The remaining 220 articles were examined directly by checking the article title and the abstract to ensure that they complied with the inclusion criteria; 48 articles were removed after that check (e.g., they did not deal with as main aim of the environmental assessment of food packaging but also considered the themes of packaged food, new environmentally friendly food packaging, and food waste). This left a total of 172 studies that matched all the inclusion criteria, and thus they constitute the object of our analysis.

4 Review Results

The first article dealing with the themes of environmental assessment of food packaging was authored by the FDA association in 1990 (Hoffman and Nowell 1990). After this first publication, which was aimed at introducing the theme of sustainability of food packaging, two other works by Kooijman (1993, 1994) dealt with the theme of the environmental assessment of food packaging. Starting from these three works onward, the increasing number of articles published shows the increasing interest in this issue with the passage of time. As reported in Table 1 and Fig. 2, the number of articles published per year has consistently increased over time.

On the basis of this preliminary analysis, we analysed the sources of the reviewed articles. The 172 articles retrieved were all published on scientific journals, in particular, *International Journal of Life Cycle Assessment* (27 articles), *Journal of Cleaner Production* (22 articles), *Packaging Technology and Science* (9 articles), and *Resources, Conservation and Recycling* (7 articles) emerged as being the journals that published most of the articles included in the review. It is also interesting to note that several studies related to the environmental assessment of food packaging were published in international journals, which fall among the waste-management disciplines of the Scopus classification (e.g., *Waste Management, Waste Management and Research*). Overall, 11 of the 172 articles included in the review were published in these two journals. This is in line with the recognized importance of food packaging in the field of waste management as previously described.

As far as the specific issues are concerned, four main topics could be identified among the reviewed articles:

Topic 1. General aspect about the environmental impact of food-packaging use and disposal (e.g., municipal waste, recycling activity, or other use)
Topic 2. Environmental assessment of packaged food including the contribution of packaging

Table 1 Number of articles on environmental assessment of food packaging or related issues per year

Year	No. of articles	Year	No. of articles	Year	No. of articles	Year	No. of articles
1990	1	1997	0	2004	5	2011	18
1991	0	1998	2	2005	2	2012	17
1992	0	1999	1	2006	4	2013	34
1993	1	2000	2	2007	6	2014	28
1994	1	2001	2	2008	13	2015	12
1995	1	2002	2	2009	7		
1996	0	2003	3	2010	10		

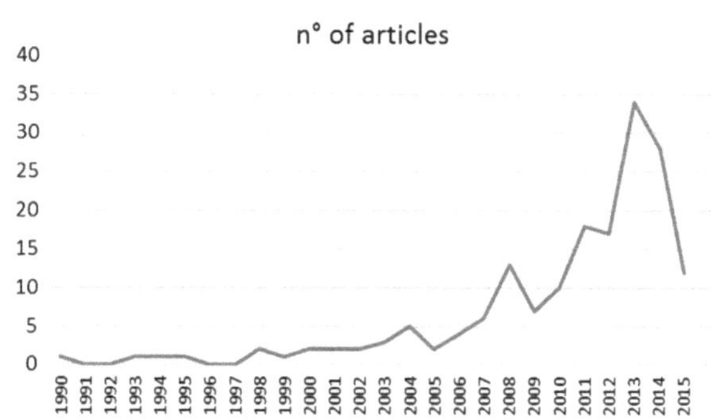

Fig. 2 Trend of numbers of articles on the environmental assessment of food packaging or related issues per year

Topic 3. Comparison of the environmental impact of several packaging systems and materials suitable for a specific food
Topic 4. New treatments or materials able to reduce the environmental impact of packaging (e.g., using biomaterials or enhanced materials and technology able to decrease food waste).

The 172 reviewed articles are classified in Table 2 based on this characterization.

Table 2 Classification of reviewed articles based on specific issues

Topic	Articles
1	Alvarenga et al. (2012), Angellier-Coussy et al. (2013), Armel et al. (2011), Arvanitoyannis and Bosnea (2001), Azapagic (2010), Barlow and Morgan (2013), Bevilacqua et al. (2008), Bugusu and Bryant (2006), Card et al. (2011), Coltro and Duarte (2013), Darlington et al. (2009), Detzel and Mönckert (2009), Dobon et al. (2011a), Dobon et al. (2011b), Edjabou et al. (2015), Edwards and Mercer (2012), Flegal et al. (2013), García-Arca et al. (2014),Gentil et al. (2011), Gentry and Shah (2004), Giugliano et al. (2011), Grizzetti et al. (2013), Grönman et al. (2013), Grosso et al. (2012), Heller and Keoleian (2003), Hoffmann and Nowell (1990), Hyde et al. (2003), Infante Amate and González De Molina (2013), Jayaraman et al. (2011), Jones (2002), Jungbluth et al. (2000), Kim et al. (2004), Kooijman (1993), Kooijman Jan (1994), Kroyer GTh (1995), Lathrop and Centner (1998), Lea and Worsley (2008), Lee et al. (2014), Li et al.(2013), Lorber et al. (2015), MacRae et al. (2013), Marsh and Bugusu (2007), Maxime et al. (2006), Meier and Christen (2013), Mena et al. (2014), Nichols et al. (2011), Oki and Sasaki (2000), Pimentel et al. (2008), Russell DAM (2014), Salhofer et al. (2008), Sanyé et al. (2012), Scipioni et al. (2013), Singh et al. (2014), Svanes et al. (2010), Tobler et al. (2011a), Van Passel (2013), Vandermeersch et al. (2014), Wan (2011), Wikström (2014), Williams et al. (2012), Xue and Landis (2010), Yano et al. (2014), Zampori and Dotelli (2014), Zhang and Wen (2014)
2	Amienyo et al. (2013), Andersson and Ohlsson (1999), Andersson et al. (1998), Bengtsson and Seddon (2013), Bevilacqua et al. (2007), Büsser and Jungbluth (2009), Calderón et al. (2010), Cellura et al. *(2012), Cordella et al. (2008), Davis and Sonesson (2008), Del Borghi et al. (2014), Espinoza-Orias et al. (2011), Flysjö (2011), Fusi et al. (2014), González-García et al. (2011), González-García et al. (2013a), Goyal et al. (2012), Hanssen (2007), Høgaas Eide (2002), Hospido et al. (2005), Hospido et al. (2006), Iribarren et al. (2010), Karakaya and Özilgen (2011), Kendall et al. (2013), Keoleian et al. (2004), Kim et al. (2013), Manfredi and Vignali (2015), Manzini et al. (2014), Marletto and Sillig (2014), Mourad et al. (2008), Nilsson et al. (2010), Oglethorpe (2009), Pardo and Zufía (2012), Pattara et al. (2012), Robertson et al. (2014), Röös et al. (2011), Roy et al. (2008), Roy et al. (2012), Sanjuán et al. (2014), Sanyé-Mengual et al. (2013), Schmidt Rivera et al. (2014), Sonesson and Berlin (2003), Talve (2001), Tanner (2006), Teixeira et al. (2013), Tobler et al. (2011b), Vázquez-Rowe et al. (2012), Vázquez-Rowe et al. (2013)
3	Accorsi et al. (2014), Accorsi et al. (2015), Albrecht et al. (2013), Azadnia et al. (2015), Banar and Çokaygil (2009), Bertoluci et al. (2014), Bø et al. (2013), Davis and Sonesson (2008), De Monte et al. (2005), Foolmaun and Ramjeeawon (2012), Humbert et al. (2009), Manfredi et al. (2015), Meneses et al. (2012), Poovarodom et al. (2012), Raheem (2013), Romero-Hernández et al. (2009), Rujnić-Sokele (2011), Silvenius et al. (2014), Siracusa et al. (2014), Toniolo et al. (2013), Von Falkenstein et al. (2010), Williams and Wikström (2011)

(continued)

Table 2 (Continued)

Topic	Articles
4	Auras et al. (2004), Bhat et al. (2013), Blanco and Siracusa (2013), Bohlmann (2004), Bugnicourt et al. (2013), Chen et al. (2014), Cheng et al. (2010), Coltelli et al. (2008), Cruz-Romero and Kerry (2008), El-Hadi (2014), Gebbink et al. (2013), González-García et al. (2013b), Hermann et al. (2010), Jamshidian et al. (2010), Juodeikiene et al. (2015), Kale et al. (2007a), Kale et al. (2007b), Leceta et al. (2013a), Leceta et al. (2013b), Leceta et al. (2015), Majeed et al. (2013), Manfredi and Vignali (2014), Mitrano et al. (2015), Montanari et al. (2014), Râpă et al. (2013), Reig et al. (2014), Rossi et al. (2015), Shen and Patel (2008), Siracusa et al. (2008), Souza et al. (2013), Tawakkal et al. (2014), Varžinskas et al. (2012), Vidal et al. (2007), Williams et al. (2008), Xu et al. (2015a), Xu et al. (2015b), Zhang et al. (2014)

As can be seen by regarding the references presented in Table 2, the majority of the articles deal with the theme of general aspects of environmental impact of food-packaging use and disposal, and this theme appears as being uniform distributed along the time span of the considered review. Following the trend shown in Fig. 1, this first topic (no. 1) in the last 5 years showed a larger number of works per year than in the past. However, when comparing the number of articles with that of overall articles per year, the ratio tends to briefly decrease (Table 3).

Regarding articles addressing topic 2, they began to appear later than those on topic 1, but in the last 10 years their number has tended to become similar to those addressing topic 1 (Fig. 3 and Table 3). This could be due to the increased number of packaged food products sold worldwide as well as the attention paid to the environmental impacts of packaging materials and technologies compared with impacts associated with food production and processing.

Based on this assumption, articles addressing topic 3 appear to be a consequence of research into the assessment of a specific packaged food product in order to find all possible solutions to decreasing its environmental impact. When investigating packaging materials and technologies, many researchers have proposed comparative solutions during the past 8 years to determine the most

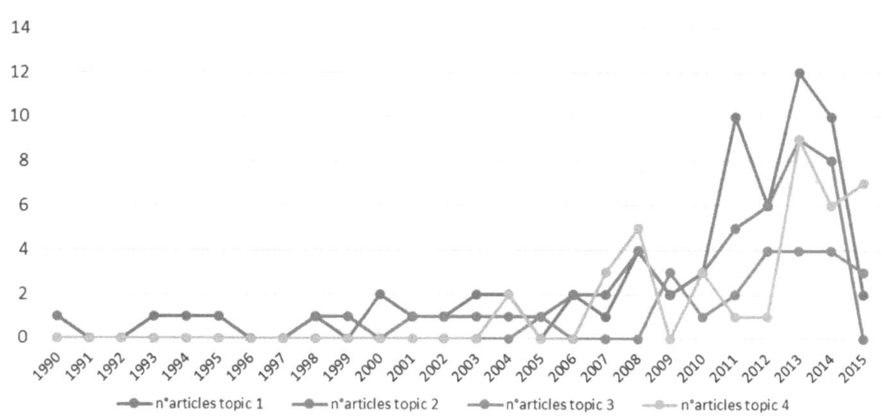

Fig. 3 Trends of articles' topics on the environmental impact of food packaging systems

Table 3 Distribution throughout 4 years of the topics of reviewed articles and the ratio between the number of articles for each topic and all reviewed articles published each year

Years	No. of articles	No. of articles topic 1	Topic 1/all	No. of articles topic 2	Topic 2/all	No. of articles topic 3	Topic 3/all	No. of articles topic 4	Topic 4/all
1990	1	1	1		0		0		0
1991	0		–		–		–		–
1992	0		–		–		–		–
1993	1	1	1		0		0		0
1994	1	1	1		0		0		0
1995	1	1	1		0		0		0
1996	0		–		–		–		–
1997	0		–		–		–		–
1998	2	1	0.5	1	0.5		0		0
1999	1		0	1	1		0		0
2000	2	2	1		0		0		0
2001	2	1	0.5	1	0.5		0		0
2002	2	1	0.5	1	0.5		0		0
2003	3	2	0.666	1	0.333		0		0
2004	5	2	0.4	1	0.2		0	2	0.4
2005	2		0	1	0.5	1	0.5		0
2006	4	2	0.5	2	0.5		0		0
2007	6	1	0.166	2	0.333		0	3	0.5
2008	13	4	0.307	4	0.307		0	5	0.384
2009	7	2	0.285	2	0.285	3	0.428		0
2010	10	3	0.3	3	0.3	1	0.1	3	0.3
2011	18	10	0.555	5	0.277	2	0.111	1	0.055
2012	17	6	0.353	6	0.353	4	0.235	1	0.058
2013	34	12	0.353	9	0.264	4	0.117	9	0.265
2014	28	10	0.357	8	0.285	4	0.142	6	0.214
2015	12	2	0.166		0	3	0.25	7	0.583
Total	172	65	0.378	48	0.279	22	0.128	37	0.215

environmentally friendly packaging. Among the 22 articles we selected regarding this topic, most of them were aimed at comparing only packaging materials; however, especially in the last 5 years, scientific research has been aimed at both materials and technologies to understand how packaging technology could play a fundamental role in reduction of the environmental impact of packaged food.

Another interesting field of interest is oriented toward the definition of new environmental friendly packaging materials, which could fall under the categories of biomaterials and recycled packaging materials for food products. Several studies have been performed by chemical scientists to find new materials able to satisfy the market requests in terms of appearance, mechanical and gas-barrier performance, and low environmental impact. However, many studies have been performed starting from 2003 that reveal a different type of evaluation mainly due to the different system boundaries and evaluation methods adopted.

As part of the proposed classification of the four topics, it has been interesting to see the evolution (when analysing the four topics) of the consideration of the impact of food-packaging technology on the amount of food waste generated. The quantity of food waste that could be reduced by means of innovative packaging technology (e.g., modified atmosphere packaging [MA]P) or active packaging solutions (e.g., oxygen scavengers, gas emitters, or antimicrobial coatings) could be relevant. It has been demonstrated, in fact, that the avoided impact of reduced food waste is considerable greater than the impacts generated by their manufacturing and the application of the new technology.

4.1 Detailed Characteristics of the Review Studies

As far as some details of the reviewed studies are concerned, it could be interesting to evaluate (where these data are reported) the following:

1. the country has been analysed
2. the functional unit assumed
3. the system boundaries assumed

Based on the high number of the reviewed studies, the main results are summarized here as follows.

The geographic area analyzed is evident because these types of studies often concern European countries. Italy, Spain, Switzerland, France, and northern European countries (e.g., Sweden, Holland, Denmark, United Kingdom, and Germany) have more times than not been cited as the survey area. Studies in the United States are often published but less so than the sum of studies concerning European countries. In addition, analyses performed in the United States have mainly been dedicated to a general evaluation of the packaging system and less concentrated on to a specific food product or packaging than is found in Europe. In the last 2 years, China has emerged as one of the countries most interested in this kind of environmental assessment. Based on this evaluation and analysing the distribution of visualized locations for some of the surveyed studies (using, for example, the Elsevier Dashboard or the ResearchGate social network), one could expect an increasing number of works coming from Asian countries in the following years. This could be supported by the fact that, as previous investigations have shown, that the number of works on the environmental assessment of food-packaging systems has been constantly increasing in the last 5 years.

It is difficult to determine categories of functional unit because they are extremely variable from case to case. As reported before, in the case of topic 2, the functional unit often reflects a specific packaged food, for example, the analysed quantities vary from <1 kg or 1 l (often exactly 1 kg or 1 l) to >1000 kg or even to the annual production of a specific country area [e.g., the total annual production of carbonated drinks in the UK (Amienyo et al. 2013)]. Other times the functional unit is related directly to a quantity of packaging often the identification of a weight (e.g., 1 kg of packaging materials or 1 kg of the final packaging including, i.e., several materials). This is the case of many articles concerning topic 3 where a comparison between several solutions has been performed comparing the same quantity of packaging materials. However, correctly, other times comparison between packaging systems or materials in topic 3 has been performed using different quantities of packaging materials because this approach is able to solve the problem of packaging food products with different weight (e.g., for some beverages a 330-g glass bottle could be equivalent to a 40-g PET bottle or a 60-g PE bottle).

The same variability is evident regarding definition of the system boundaries. As shown in Fig. 1, several choices are possible based on the main aim of the study. In some cases, e.g., only the disposal phase is analysed (e.g., Foolmaun and Ramjeeawon 2012), especially if the interest of the journal is directed toward waste management or similar issues. Most frequent is obviously analysis of the primary packaging phase or the manufacturing phase of the primary packaging materials. Analysis of food production, starting not from the agricultural phase but considering the food ingredients at the beginning of the food industry process, is a vey frequent topic. As stated previously in the presentation of the topics, recently the food waste has been studied due to its recognised impact on the environmental burdens. Finally, some studies, published mainly in some production research, logistics, or supply chain journals, have focused on the impact of transporting packaging materials along the entire food supply chain, thus showing that this phase could be relevant in the evaluation of the environmental assessment of a packaged food product (Accorsi et al. 2014, 2015; Albrect et al. 2013).

Other possible classifications of the reviewed studies could be performed evaluating, for example, as done by Ng et al. (2013) in another review of LCA studies regarding diapers, i.e., the variants of the packaging system under consideration by each particular study, the assumption of the studies, and the major conclusions arising from each study. However, the author of the present study, due to the wide spectrum of this review analysis, considers these further in-depth analyses as not extremely useful to understanding the direction of the research field on LCA of food-packaging systems. As described in the following section, this could be interesting for some limited issues of interest inside the food-packaging sector, and it could be dealt with in future research, for example, by the author's research group.

5 Conclusions and Future Research

Based on the performed review, evolution in the assessment of packaging systems is evident. Starting mainly from an evaluation of the environmental impact of disposed packaging materials, scientific community moved toward an evaluation of the impact of packaging materials and technology compared with the impact associated with the food product contained by each packaging. Several works are still being performed on this issue with the main aim to explore some part of the food supply chain not that has not yet studied. With the aim of comparing several packaging solutions, some comparative LCAs have been performed in the last years mainly focusing on the impact of packaging materials and only a few times also considering the impact of packaging technology and machineries. In this latter sector, the manufacturing phase of packaging equipment has never been considered because of the limited impact due to the long life time of packaging equipment. The evolution of biobased and recycled packaging materials also gave rise in the past 12 years to research aiming at demonstrating the limited impact of these new materials, which are in contact with food.

Regarding the method of analysis, the main evolution is represented by the extension of the analysed system boundaries; actually, all the phases of a specific food supply chain are considered excluded. Eventually, in a comparative analysis of packaging systems only the phases not affected by a change of packaging technology will be considered. Regarding the impact-assessment method, authors chose the best one based on the main findings they want to obtain (considering, for example, sometimes a toxicity aspect or not). Among the most adopted methods included are the CML 2001 (Guinée et al. 2001), the ReCiPe 2008 (Goedkoop et al. 2009), the ILCD 2011 (European Commission's Joint Research Centre, 2010), and the Impact 2002+ (Jolliet et al. 2003).

Finally, as far as future researches in this field are concerned, the environmental evaluation of new intelligent and effective packaging design that is able to extend the shelf life of packaged food products, even also after their opened, appears very promising. Thanks to the extension of the shelf life of the packaged food product, this kind of applications allows reduction of the quantity of food wasted, thus allowing substantial benefits in the reduction of environmental impacts. In the majority of the cases, in fact, the decreased impact due to avoided food waste is greater than the impact generated by the introduction of this new technology. As suggested by several authors, much research should also be performed to better survey the possibility of their application in several food supply chains.

As a conclusion of the presented review, innovations in food packaging systems can be considered as one of the main topic to study with in the next few years to reduce the impact of food product packaging especially regarding food waste.

References

Accorsi R, Cascini A, Cholette S, Manzini R, Mora C (2014) Economic and environmental assessment of reusable plastic containers: a food catering supply chain case study. Int J Prod Econ 152:88–101

Accorsi R, Versari L, Manzini R (2015) Glass vs. plastic: life cycle assessment of extra-virgin olive oil bottles across global supply chains. Sustainability (Switzerland), 7(3):2818–2840

Albrecht S, Brandstetter P, Beck T, Fullana-I-Palmer P, Grönman K, Baitz M, Deimling S, Sandilands J, Fischer M (2013) An extended life cycle analysis of packaging systems for fruit and vegetable transport in Europe. Int J Life Cycle Assess 18(8):1549–1567

Alderson P, Green S, Higgins JPT (eds) (2004) Cochrane reviewers' handbook 4.2.2 (updated March 2004). Available at http://www.cochrane.org/resources/handbook/hbook.htm. (Accessed June 2015)
Almeida CMVB, Rodrigues AJM, Bonilla SH, Giannetti BF (2010) Emergy as a tool for Ecodesign: evaluating materials selection for beverage packages in Brazil. J Clean Prod 18(1):32–43
Alvarenga RAFD, Galindro BM, Helpa CDF, Soares SR (2012) The recycling of oyster shells: an environmental analysis using Life Cycle Assessment. J Environ Manage 106:102–109
Amienyo D, Gujba H, Stichnothe H, Azapagic A (2013) Life cycle environmental impacts of carbonated soft drinks. Int J Life Cycle Assess 18(1):77–92
Andersson K, Ohlsson T (1999) Life cycle assessment of bread produced on different scales. Int J Life Cycle Assess 4(1):25–40
Andersson K, Ohlsson T, Olsson P (1998) Screening life cycle assessment (LCA) of tomato ketchup: a case study. J Clean Prod 6(3–4):277–288
Angellier-Coussy H, Guillard V, Guillaume C, Gontard N (2013) Role of packaging in the smorgasbord of action for sustainable food consumption. Agro Food Ind Hi-Tech 24(3):15–19
Armel KC, Yan K, Todd A, Robinson TN (2011) The Stanford Climate Change Behavior Survey (SCCBS): assessing greenhouse gas emissions-related behaviors in individuals and populations. Clim Change 109(3–4):671–694
Arvanitoyannis IS, Bosnea LA (2001) Recycling of polymeric materials used for food packaging: current status and perspectives. Food Rev Int 17(3):291–346
Auras R, Harte B, Selke S (2004) An overview of polylactides as packaging materials. Macromol Biosci 4(9):835–864
Azadnia AH, Saman MZM, Wong KY (2015) Sustainable supplier selection and order lot-sizing: an integrated multi-objective decision-making process. Int J Prod Res 53(2):383–408
Azapagic A (2010) Life cycle assessment as a tool for sustainable management of ecosystem services. Issues Environ Sci Technol 30:140–168
Banar M, Çokaygil Z (2009) A life cycle comparison of alternative cheese packages. Clean: Soil Air Water 37(2):136–141
Barlow CY, Morgan DC (2013) Polymer film packaging for food: an environmental assessment. Resour Conserv Recycl 78:74–80
Bengtsson J, Seddon J (2013) Cradle to retailer or quick service restaurant gate life cycle assessment of chicken products in Australia. J Clean Prod 41:291–300
Bertoluci G, Leroy Y, Olsson A (2014) Exploring the environmental impacts of olive packaging solutions for the European food market. J Clean Prod 64:234–243
Bevilacqua M, Braglia M, Carmignani G, Zammori FA (2007) Life cycle assessment of pasta production in Italy. J Food Qual 30(6):932–952
Bevilacqua M, Ciarapica FE, Giacchetta G (2008) Design for environment as a tool for the development of a sustainable supply chain. Int J Sustain Eng 1(3):188–201
Bhat R, Abdullah N, Din RH, Tay G-S (2013) Producing novel sago starch based food packaging films by incorporating lignin isolated from oil palm black liquor waste. J Food Eng 119(4):707–713
Blanco I, Siracusa V (2013) Kinetic study of the thermal and thermo-oxidative degradations of polylactide-modified films for food packaging. J Therm Anal Calorim 112(3):1171–1177
Bø E, Hammervoll T, Tvedt K (2013) Environmental impact of refillable vs. non-refillable plastic beverage bottles in Norway. Int J Environ Sustain Dev 12(4):379–395
Bohlmann GM (2004) Biodegradable packaging life-cycle assessment. Environ Prog 23(4):342–346
Bugnicourt E, Schmid M, Mc Nerney O, Wildner J, Smykala L, Lazzeri A, Cinelli P (2013) Processing and validation of whey-protein-coated films and laminates at semi-industrial scale as novel recyclable food packaging materials with excellent barrier properties. Advances in Materials Science and Engineering, Article ID 496207, 10 p. doi:http://dx.doi.org/10.1155/2013/496207
Bugusu B, Bryant C (2006) Defining the future of food packaging. Food Technol 60(12):38–42
Büsser S, Jungbluth N (2009) The role of flexible packaging in the life cycle of coffee and butter. Int J Life Cycle Assess 14(1):80–91

Calderón LA, Iglesias L, Laca A, Herrero M, Díaz M (2010) The utility of Life Cycle Assessment in the ready meal food industry. Resour Conserv Recycl 54(12):1196–1207
Card JW, Jonaitis TS, Tafazoli S, Magnuson BA (2011) An appraisal of the published literature on the safety and toxicity of food-related nanomaterials. Crit Rev Toxicol 41(1):22–51
Cellura M, Longo S, Mistretta M (2012) Life Cycle Assessment (LCA) of protected crops: an Italian case study. J Clean Prod 28:56–62
Chen H, Seiber JN, Hotze M (2014) ACS select on nanotechnology in food and agriculture: a perspective on implications and applications. J Agric Food Chem 62(6):1209–1212
Cheng X, Shi H, Adams CD, Ma Y (2010) Assessment of metal contaminations leaching out from recycling plastic bottles upon treatments. Environ Sci Pollut Res 17(7):1323–1330
Coltelli M-B, Maggiore ID, Bertoldo M, Signori F, Bronco S, Ciardelli F (2008) Poly(lactic acid) properties as a consequence of poly(butylene adipate-co-terephthalate) blending and acetyl tributyl citrate plasticization. J Appl Polym Sci 110(2):1250–1262
Coltro L, Duarte LC (2013) Flexible plastic packaging recycling: the contribution of the correct identification [Reciclagem de embalagens plásticas flexíveis: Contribuição da identificação correta]. Polimeros 23(1):128–134
Cook DJ, Mulrow CD, Haynes RB (1997) Systematic reviews: synthesis of best evidence for clinical decisions. Ann Intern Med 126(5):379–380
Cordella M, Tugnoli A, Spadoni G, Santarelli F, Zangrando T (2008) LCA of an Italian lager beer. Int J Life Cycle Assess 13(2):133–139
Cruz-Romero M, Kerry JP (2008) Crop-based biodegradable packaging and its environmental implications. CAB Rev: Perspect Agric Vet Sci Nutr Nat Resour 3(074):1–25
Darlington R, Staikos T, Rahimifard S (2009) Analytical methods for waste minimisation in the convenience food industry. Waste Manag 29(4):1274–1281
Davies GF, Man CMD, Andrews SD, Paurine A, Hutchins MG, Maidment GG (2012) Potential life cycle carbon savings with low emissivity packaging for refrigerated food on display. J Food Eng 109(2):202–208
Davis J, Sonesson U (2008) Life cycle assessment of integrated food chains—a Swedish case study of two chicken meals. Int J Life Cycle Assess 13(7):574–584
De Monte M, Padoano E, Pozzetto D (2005) Alternative coffee packaging: an analysis from a life cycle point of view. J Food Eng 66(4):405–411
Del Borghi A, Gallo M, Strazza C, Del Borghi M (2014) An evaluation of environmental sustainability in the food industry through Life Cycle Assessment: the case study of tomato products supply chain. J Clean Prod 78:121–130
Denafas G, Ruzgas T, Martuzevicius D, Shmarin S, Hoffmann M, Mykhaylenko V, Ogorodnik S, Romanov M, Neguliaeva E, Chusov A, Turkadze T, Bochoidze I, Ludwig C (2014) Seasonal variation of municipal solid waste generation and composition in four East European cities. Resour Conserv Recycl 89:22–30
Detzel A, Mönckert J (2009) Environmental evaluation of aluminium cans for beverages in the German context. Int J Life Cycle Assess 14(1):70–79
Dobon A, Cordero P, Kreft F, Østergaard SR, Antvorskov H, Robertsson M, Smolander M, Hortal M (2011a) The sustainability of communicative packaging concepts in the food supply chain. A case study: Part 2. Life cycle costing and sustainability assessment. Int J Life Cycle Assess 16(6):537–547
Dobon A, Cordero P, Kreft F, Østergaard SR, Robertsson M, Smolander M, Hortal M (2011b) The sustainability of communicative packaging concepts in the food supply chain. A case study: Part 1. Life cycle assessment. Int J Life Cycle Assess 16(2):168–177
Edjabou ME, Jensen MB, Götze R, Pivnenko K, Petersen C, Scheutz C, Astrup TF (2015) Municipal solid waste composition: sampling methodology, statistical analyses, and case study evaluation. Waste Manag 36:12–23
Edwards F, Mercer D (2012) Food waste in Australia: the freegan response. Sociol Rev 60(2):174–191
El-Hadi AM (2014) Development of novel biopolymer blends based on poly(L-lactic acid), poly((R)-3-hydroxybutyrate), and plasticizer. Polym Eng Sci 54(6):1394–1402
Espinoza-Orias N, Stichnothe H, Azapagic A (2011) The carbon footprint of bread. Int J Life Cycle Assess 16(4):351–365

European Commission's Joint Research Centre (2010) International Reference Life Cycle Data System (ILCD) Handbook. In: General Guidance for Life Cycle Assessment. EC-JRC, Luxembourg

Eurostat (2014) Packaging waste. Available on line: http://appsso.eurostat.ec.europa.eu

Flegal AR, Gallon C, Ganguli PM, Conaway CH (2013) All the lead in China. Crit Rev Environ Sci Technol 43(17):1869–1944

Flysjö A (2011) Potential for improving the carbon footprint of butter and blend products. J Dairy Sci 94(12):5833–5841

Foolmaun RK, Ramjeeawon T (2012) Disposal of post-consumer polyethylene terephthalate (PET) bottles: comparison of five disposal alternatives in the small island state of Mauritius using a life cycle assessment tool. Environ Technol 33(5):563–572

Fusi A, Guidetti R, Benedetto G (2014) Delving into the environmental aspect of a Sardinian white wine: from partial to total life cycle assessment. Sci Total Environ 472:989–1000

García-Arca J, Carlos Prado-Prado J, Trinidad Gonzalez-Portela Garrido A (2014) Packaging logistics: promoting sustainable efficiency in supply chains. Int J Phys Distrib Logistics Manag 44(4):325–346

Gebbink WA, Ullah S, Sandblom O, Berger U (2013) Polyfluoroalkyl phosphate esters and perfluoroalkyl carboxylic acids in target food samples and packaging-method development and screening. Environ Sci Pollut Res 20(11):7949–7958

Gentil EC, Gallo D, Christensen TH (2011) Environmental evaluation of municipal waste prevention. Waste Manag 31(12):2371–2379

Gentry AE, Shah MN (2004) A life-cycle approach to assuring compliance of the production system. Am Pharm Outsourcing 5(1):8–14

Giugliano M, Cernuschi S, Grosso M, Rigamonti L (2011) Material and energy recovery in integrated waste management systems. An evaluation based on life cycle assessment. Waste Manage 31(9–10):2092–2101

Goedkoop M, Heijungs R, Huijbregts M, De Schryver A, Struijs J, Zelm R (2009) ReCiPe 2008. A Life Cycle Impact Assessment Method Which Comprises Harmonised Category Indicators at the Midpoint and the Endpoint Level

Gomez G, Meneses M, Ballinas L, Castells F (2008) Characterization of urban solid waste in Chihuahua, Mexico. Waste Manage 28(12):2465–2471

González-García S, Silva FJ, Moreira MT, Pascual RC, Lozano RG, Gabarrell X, Pons JRI, Feijoo G (2011) Combined application of LCA and eco-design for the sustainable production of wood boxes for wine bottles storage. Int J Life Cycle Assess 16(3):224–237

González-García S, Castanheira EG, Dias AC, Arroja L (2013a) Environmental life cycle assessment of a dairy product: the yoghurt. Int J Life Cycle Assess 18(4):796–811

González-García S, Castanheira TG, Dias AC, Arroja L (2013b) Using Life Cycle Assessment methodology to assess UHT milk production in Portugal. Sci Total Environ 442:225–234

Goyal SK, Jogdand SV, Agrawal AK (2012) Energy use pattern in rice milling industries—a critical appraisal. J Food Sci Technol 51(11):2907–2916

Grizzetti B, Pretato U, Lassaletta L, Billen G, Garnier J (2013) The contribution of food waste to global and European nitrogen pollution. Environ Sci Policy 33:186–195

Grönman K, Soukka R, Järvi-Kääriäinen T, Katajajuuri J-M, Kuisma M, Koivupuro H-K, Ollila M, Pitkänen M, Miettinen O, Silvenius F, Thun R, Wessman H, Linnanen L (2013) Framework for sustainable food packaging design. Packag Technol Sci 26:187–200

Grosso M, Nava C, Testori R, Rigamonti L, Vigano F (2012) The implementation of anaerobic digestion of food waste in a highly populated urban area: an LCA evaluation. Waste Manage Res 30(9)(1):78–87

Guinée JB (2001). Life Cycle Assessment: an Operational Guide to the ISO Standards. Ministry of Housing. Spatial Planning and Environment (VROM) and Centre of Environmental Science (CML), The Netherlands

Hanssen OJ, Rukke E-O, Saugen B, Kolstad J, Hafrom P, Von Krogh L, Raadal HL, Rønning A, Wigum KS (2007) The environmental effectiveness of the beverage sector in Norway in a factor 10 perspective. Int J Life Cycle Assess 12(4):257–265

Heller MC, Keoleian GA (2003) Assessing the sustainability of the US food system: a life cycle perspective. Agric Syst 76(3):1007–1041
Hermann BG, Blok K, Patel MK (2010) Twisting biomaterials around your little finger: environmental impacts of bio-based wrappings. Int J Life Cycle Assess 15(4):346–358
Hoffmann BL, Nowell LH (1990) The FDA's environmental impact analysis requirements: food packaging and solid waste management. Food Drug Cosmet Law J 45(6):615–621
Høgaas Eide M (2002) Life cycle assessment (LCA) of industrial milk production. Int J Life Cycle Assess 7(2):115–126
Hospido A, Moreira MT, Feijoo G (2005) Environmental analysis of beer production. Int J Agric Resour Gov Ecol 4(2):152–162
Hospido A, Vazquez ME, Cuevas A, Feijoo G, Moreira MT (2006) Environmental assessment of canned tuna manufacture with a life-cycle perspective. Resour Conserv Recycl 47(1):56–72
Humbert S, Rossi V, Margni M, Jolliet O, Loerincik Y (2009) Life cycle assessment of two baby food packaging alternatives: glass jars vs. plastic pots. Int J Life Cycle Assess 14(2):95–106
Hyde K, Miller L, Smith A, Tolliday J (2003) Minimising waste in the food and drink sector: using the business club approach to facilitate training and organisational development. J Environ Manage 67(4):327–338
Infante Amate J, González De Molina M (2013) 'Sustainable de-growth' in agriculture and food: an agro-ecological perspective on Spain's agri-food system (year 2000). J Clean Prod 38:27–35
Iribarren D, Hospido A, Moreira MT, Feijoo G (2010) Carbon footprint of canned mussels from a business-to-consumer approach. A starting point for mussel processors and policy makers. Environ Sci Policy 13(6):509–521
ISO 14040 (2006) Environmental management—Life Cycle Assessment-Principles and Framework. International Organisation for Standardisation. Geneva, (CH)
Jamshidian M, Tehrany EA, Imran M, Jacquot M, Desobry S (2010) Poly-Lactic Acid: production, applications, nanocomposites, and release studies. Compr Rev Food Sci Food Saf 9(5):552–571
Jayaraman K, Haron H, Sung GB, Lin SK (2011) Consumer reflections on the usage of plastic bags to parcel hot edible items: An empirical study in Malaysia. J Clean Prod 19(13):1527–1535
Jolliet O, Margni M, Charles R, Humbert S, Payet J, Rebitzer G and Rosenbaum R (2003) IMPACT 2002+: A New Life Cycle Impact Assessment Methodology. Int J Life Cycle Assess 8(6):324–330
Jones A (2002) An environmental assessment of food supply chains: a case study on dessert apples. Environ Manage 30(4):560–576
Jungbluth N, Tietje O, Scholz RW (2000) Food purchases: impacts from the consumers' point of view investigated with a modular LCA. Int J Life Cycle Assess 5(3):134–142
Juodeikiene G, Vidmantiene D, Basinskiene L, Cernauskas D, Bartkiene E, Cizeikiene D (2015) Green metrics for sustainability of biobased lactic acid from starchy biomass vs chemical synthesis. Catal Today 239:11–16
Kale G, Auras R, Singh SP (2007a) Comparison of the degradability of poly(lactide) packages in composting and ambient exposure conditions. Packag Technol Sci 20(1):49–70
Kale G, Kijchavengkul T, Auras R, Rubino M, Selke SE, Singh SP (2007b) Compostability of bioplastic packaging materials: an overview. Macromol Biosci 7(3):255–277
Karakaya A, Özilgen M (2011) Energy utilization and carbon dioxide emission in the fresh, paste, whole-peeled, diced, and juiced tomato production processes. Energy 36(8):5101–5110
Kendall A, Yuan J, Brodt SB (2013) Carbon footprint and air emissions inventories for US honey production: case studies. Int J Life Cycle Assess 18(2):392–400
Keoleian GA, Phipps AW, Dritz T, Brachfeld D (2004) Life cycle environmental performance and improvement of a yogurt product delivery system. Packag Technol Sci 17(2):85–103
Kim D, Thoma G, Nutter D, Milani F, Ulrich R, Norris G (2013) Life cycle assessment of cheese and whey production in the USA. Int J Life Cycle Assess 18(5):1019–1035

Kim HJ, Jo BM, Hwang YW, Park KH (2004) Environmental impact evaluation of virgin pulp using life cycle assessment methodology. Palpu Chongi Gisul/J Korea Tech Assoc Pulp Paper Ind 36(1):49–60

Kooijman JM (1993) Environmental assessment of packaging: sense and sensibility. Environ Manage 17(5):575–586

Kooijman Jan M (1994) Environmental assessment of food packaging: impact and improvement. Packag Technol Sci 7(3):111–121

Kroyer GTh (1995) Impact of food processing on the environment-an overview. LWT: Food Sci Technol 28(6):547–552

Lathrop KW, Centner TJ (1998) Eco-labeling and ISO 14000: an analysis of US regulatory systems and issues concerning adoption of type II standards. Environ Manage 22(2):163–172

Lea E, Worsley A (2008) Australian consumers' food-related environmental beliefs and behaviours. Appetite 50(2–3):207–214

Leceta I, Guerrero P, Cabezudo S, De La Caba K (2013a) Environmental assessment of chitosan-based films. J Clean Prod 41:312–318

Leceta I, Guerrero P, Ibarburu I, Dueñas MT, De La Caba K (2013b) Characterization and antimicrobial analysis of chitosan-based films. J Food Eng 116(4):889–899

Leceta I, Uranga J, Arana P, Cabezudo S, De La Caba K, Guerrero P (2015) Valorisation of fishery industry wastes to manufacture sustainable packaging films: modelling moisture-sorption behaviour. J Clean Prod 91:36–42

Lee J, Pedersen AB, Thomsen M (2014) The influence of resource strategies on childhood phthalate exposure—The role of REACH in a zero waste society. Environ Int 73:312–322

Li L, Diederick R, Flora JRV, Berge ND (2013) Hydrothermal carbonization of food waste and associated packaging materials for energy source generation. Waste Manag 33(11):2478–2492

Lorber M, Schecter A, Paepke O, Shropshire W, Christensen K, Birnbaum L (2015) Exposure assessment of adult intake of bisphenol A (BPA) with emphasis on canned food dietary exposures. Environ Int 77:55–62

MacRae R, Cuddeford V, Young SB, Matsubuchi-Shaw M (2013) The food system and climate change: an exploration of emerging strategies to reduce GHG emissions in Canada. Agroecology Sustain Food Syst 37(8):933–963

Majeed K, Jawaid M, Hassan A, Abu Bakar A, Abdul Khalil HPS, Salema AA, Inuwa I (2013) Potential materials for food packaging from nanoclay/natural fibres filled hybrid composites. Mater Des 46:391–410

Manfredi M, Vignali G (2014) LCA of a packaged tomato puree: a comparison of environmental impacts produced by different life cycle phases. J Clean Prod 73:275–284

Manfredi M, Vignali G (2015) Comparative life cycle assessment of hot filling and aseptic packaging systems used for beverages. J Food Eng 147:39–48

Manfredi M, Fantin V, Vignali G, Gavara R (2015) Environmental assessment of antimicrobial coatings for packaged fresh milk. J Clean Prod 95:291–300

Manzini R, Accorsi R, Ayyad Z, Bendini A, Bortolini M, Gamberi M, Valli E, Toschi TG (2014) Sustainability and quality in the food supply chain. A case study of shipment of edible oils. Br Food J 116(12):2069–2090

Marletto G, Sillig C (2014) Environmental impact of Italian canned tomato logistics: national vs. regional supply chains. J Transp Geogr 34:131–141

Marsh K, Bugusu B (2007) Food packaging—roles, materials, and environmental issues: scientific status summary. J Food Sci 72(3):39–55

Maxime D, Marcotte M, Arcand Y (2006) Development of eco-efficiency indicators for the Canadian food and beverage industry. J Clean Prod 14(6–7):636–648

Meier T, Christen O (2013) Environmental impacts of dietary recommendations and dietary styles: Germany as an example. Environ Sci Technol 47(2):877–888

Mena C, Terry LA, Williams A, Ellram L (2014) Causes of waste across multi-tier supply networks: cases in the UK food sector. Int J Prod Econ 152:144–158

Meneses M, Pasqualino J, Castells F (2012) Environmental assessment of the milk life cycle: the effect of packaging selection and the variability of milk production data. J Environ Manage 107:76–83
Mitrano DM, Motellier S, Clavaguera S, Nowack B (2015) Review of nanomaterial aging and transformations through the life cycle of nano-enhanced products. Environ Int 77:132–147
Montanari A, Bolzoni L, Cigognini IM, De La Torre Carreras R (2014) Tomato bio-based lacquer: for sustainable metal packaging. Agro Food Industry Hi-Tech 25(1):50–54
Mourad AL, Garcia EEC, Vilela GB, Von Zuben F (2008) Influence of recycling rate increase of aseptic carton for long-life milk on GWP reduction. Resour Conserv Recycl 52(4):678–689
Ng FS-F, Muthu SS, Li Y, Hui PC-L (2013) A critical review on life cycle assessment studies of diapers. Crit Rev Environ Sci Technol 43(16):1795–1822
Nichols J, Kang S, Post W, Wang D, Bandaru V, Manowitz D, Zhang X, Izaurralde R (2011) HPC-EPIC for high resolution simulations of environmental and sustainability assessment. Comput Electron Agric 79(2):112–115
Nilsson K, Flysjö A, Davis J, Sim S, Unger N, Bell S (2010) Comparative life cycle assessment of margarine and butter consumed in the UK, Germany and France. Int J Life Cycle Assess 15(9):916–926
Oglethorpe D (2009) Food miles—the economic, environmental and social significance of the focus on local food. CAB Rev: Perspect Agric Vet Sci Nutr Nat Resour 4(72):1–11
Oki Y, Sasaki H (2000) Social and environmental impacts of packaging (LCA and assessment of packaging functions). Packag Technol Sci 13(2):45–53
Pardo G, Zufía J (2012) Life cycle assessment of food-preservation technologies. J Clean Prod 28:198–207
Pattara C, Raggi A, Cichelli A (2012) Life cycle assessment and carbon footprint in the wine supply-chain. Environ Manage 49(6):1247–1258
Pimentel D, Williamson S, Alexander CE, Gonzalez-Pagan O, Kontak C, Mulkey SE (2008) Reducing energy inputs in the US food system. Human Ecol 36(4):459–471
Poovarodom N, Ponnak C, Manatphrom N (2012) Comparative carbon footprint of packaging systems for tuna products. Packag Technol Sci 25(5):249–257
Raheem D (2013) Application of plastics and paper as food packaging materials—An overview. Emirates J Food Agric 25(3):177–188
Râpă M, Ghioca PN, Grosu E, Iancu L, Spurcaciu B, Grigorescu R, Trifoi A, Cincu C (2013) Development of new recycled polypropylene/styreneisoprene-styrene block copolymers composites. J Optoelectron Adv Mater 15(7–8):817–822
Reig CS, Lopez AD, Ramos MH, Cloquell Ballester VA (2014) Nanomaterials: a map for their selection in food packaging applications. Packag Technol Sci 27(11):839–866
Robertson GL (2012) Food packaging: principles and practice, 3rd edn. CRC Press, Boca Raton
Robertson K, Garnham M, Symes W (2014) Life cycle carbon footprint of the packaging and transport of New Zealand Kiwifruit. Int J Life Cycle Assess 19(10):1693–1704
Romero-Hernández O, Romero Hernández S, Muñoz D, Detta-Silveira E, Palacios-Brun A, Laguna A (2009) environmental implications and market analysis of soft drink packaging systems in Mexico. A waste management approach. Int J Life Cycle Assess 14(2):107–113
Röös E, Sundberg C, Hansson P-A (2011) Uncertainties in the carbon footprint of refined wheat products: a case study on Swedish pasta. Int J Life Cycle Assess 16(4):338–350
Rossi V, Cleeve-Edwards N, Lundquist L, Schenker U, Dubois C, Humbert S, Jolliet O (2015) Life cycle assessment of end-of-life options for two biodegradable packaging materials: sound application of the European waste hierarchy. J Clean Prod 86:132–145
Roy P, Nei D, Okadome H, Nakamura N, Orikasa T, Shiina T (2008) Life cycle inventory analysis of fresh tomato distribution systems in Japan considering the quality aspect. J Food Eng 86(2):225–233
Roy P, Nei D, Orikasa T, Xu Q, Okadome H, Nakamura N, Shiina T (2009) A review of life cycle assessment (LCA) on some food products. J Food Eng 90(1):1–10
Roy P, Orikasa T, Thammawong M, Nakamura N, Xu Q, Shiina T (2012) Life cycle of meats: an opportunity to abate the greenhouse gas emission from meat industry in Japan. J Environ Manage 93(1):218–224

Rujnić-Sokele M (2011) Plastic packaging—the best choice for the environment [Plastična ambalaža—Najbolji izbor za okoliš]. Polimeri (Zagreb) 32(2):80–82
Russell DAM (2014) Sustainable (food) packaging—an overview. Food Addit Contam Part A Chem Anal Control Expo Risk Assess 31(3):396–401
Salhofer S, Obersteiner G, Schneider F, Lebersorger S (2008) Potentials for the prevention of municipal solid waste. Waste Manage 28(2):245–259
Sanjuán N, Stoessel F, Hellweg S (2014) Closing data gaps for LCA of food products: estimating the energy demand of food processing. Environ Sci Technol 48(2):1132–1140
Sanyé E, Oliver-Solà J, Gasol CM, Farreny R, Rieradevall J, Gabarrell X (2012) Life cycle assessment of energy flow and packaging use in food purchasing. J Clean Prod 25:51–59
Sanyé-Mengual E, Cerón-Palma I, Oliver-Solá J, Montero JI, Rieradevall J (2013) Environmental analysis of the logistics of agricultural products from roof top greenhouses in mediterranean urban areas. J Sci Food Agric 93(1):100–109
Schmidt Rivera XC, Espinoza Orias N, Azapagic A (2014) Life cycle environmental impacts of convenience food: comparison of ready and home-made meals. J Clean Prod 73:294–309
Scipioni A, Niero M, Mazzi A, Manzardo A, Piubello S (2013) Significance of the use of non-renewable fossil CED as proxy indicator for screening LCA in the beverage packaging sector. Int J Life Cycle Assess 18(3):673–682
Sealey KS, Smith J (2014) Recycling for small island tourism developments: food waste composting at Sandals Emerald Bay, Exuma, Bahamas. Resour Conserv Recycl 92:25–37
Shen L, Patel MK (2008) Life cycle assessment of polysaccharide materials: a review. J Polym Environ 16(2):154–167
Silvenius F, Grönman K, Katajajuuri J-M, Soukka R, Koivupuro H-K, Virtanen Y (2014) The role of household food waste in comparing environmental impacts of packaging alternatives. Packag Technol Sci 27(4):277–292
Singh M, Marchis A, Capri E (2014) Greening, new frontiers for research and employment in the agro-food sector. Sci Total Environ 472:437–443
Siracusa V, Ingrao C, Lo Giudice A, Mbohwa C, Dalla Rosa M (2014) Environmental assessment of a multilayer polymer bag for food packaging and preservation: an LCA approach. Food Res Int 62:151–161
Siracusa V, Rocculi P, Romani S, Rosa MD (2008) Biodegradable polymers for food packaging: a review. Trends Food Sci Technol 19(12):634–643
Sonesson U, Berlin J (2003) Environmental impact of future milk supply chains in Sweden: a scenario study. J Clean Prod 11(3):253–266
Sonneveld K (2000) Role of life cycle assessment as a decision support tool for packaging. Packag Technol Sci 13(2):55–61
Souza PMS, Morales AR, Marin-Morales MA, Mei LHI (2013) PLA and montmorilonite nanocomposites: properties, biodegradation and potential toxicity. J Polym Environ 21(3):738–759
Styles D, Schoenberger H, Galvez-Martos J-L (2012) Environmental improvement of product supply chains: a review of European retailers' performance. Resour Conserv Recycl 65:57–78
Svanes E, Void M, Møller H, Pettersen MK, Larsen H, Hanssen OJ (2010) Sustainable packaging design: a holistic methodology for packaging design. Packag Technol Sci 23(3):161–175
Talve S (2001) Life cycle assessment of a basic lager beer. Int J Life Cycle Assess 6(5):293–298
Tanner C (2006) When consumers judge the environmental significance of food products: results of experimental studies [Wenn konsumentinnen und konsumenten die umweltverträglichkeit von lebensmitteln beurteilen: Ergebnisse experimenteller studien]. GAIA 15(3):215–220
Tawakkal ISMA, Cran MJ, Miltz J, Bigger SW (2014) A review of poly(lactic acid)-based materials for antimicrobial packaging. J Food Sci 79(8):1477–1490
Teixeira R, Himeno A, Gustavus L (2013) Carbon footprint of Breton pâté production: a case study. Integr Environ Assess Manage 9(4):645–651
Tobler C, Visschers VHM, Siegrist M (2011a) Eating green. Consumers' willingness to adopt ecological food consumption behaviors. Appetite 57(3):674–682

Tobler C, Visschers VHM, Siegrist M (2011b) Organic tomatoes versus canned beans: how do consumers assess the environmental friendliness of vegetables? Environ Behav 43(5):591–611
Toniolo S, Mazzi A, Niero M, Zuliani F, Scipioni A (2013) Comparative LCA to evaluate how much recycling is environmentally favourable for food packaging. Resour Conserv Recycl 77:61–68
Transfield D, Denyer D, Palminder S (2003) Towards a methodology for developing evidence-informed management knowledge by means of systematic review. Br J Manag 14:207–222
Van Passel S (2013) Food miles to assess sustainability: a revision. Sustain Dev 21(1):1–17
Vandermeersch T, Alvarenga RAF, Ragaert P, Dewulf J (2014) Environmental sustainability assessment of food waste valorization options. Resour Conserv Recycl 87:57–64
Varžinskas V, Staniškis JK, Knašyte M (2012) Decision-making support system based on LCA for aseptic packaging recycling. Waste Manage Res 30(9):931–939
Vázquez-Rowe I, Moreira MT, Feijoo G (2012) Environmental assessment of frozen common octopus (Octopus vulgaris) captured by Spanish fishing vessels in the Mauritanian EEZ. Marine Policy 36(1):180–188
Vázquez-Rowe I, Villanueva-Rey P, Mallo J, De La Cerda JJ, Moreira MT, Feijoo G (2013) Carbon footprint of a multi-ingredient seafood product from a business-to-business perspective. J Clean Prod 44:200–210
Vidal R, Martínez P, Mulet E, González R, López-Mesa B, Fowler P, Fang JM (2007) Environmental assessment of biodegradable multilayer film derived from carbohydrate polymers. J Polym Environ 15(3):159–168
Von Falkenstein E, Wellenreuther F, Detzel A (2010) LCA studies comparing beverage cartons and alternative packaging: can overall conclusions be drawn? Int J Life Cycle Assess 15(9):938–945
Wan J (2011) Report from Asia: packaging is indeed headed to be a lead sector in the Asian pharmaceutical environment, but certain challenges must first be overcome. Pharm Technol 35(12):14–16
Wever R, Vogtländer J (2013) Eco-efficient value creation: an alternative perspective on packaging and sustainability. Packag Technol Sci 26(4):229–248
Wikström F, Williams H, Verghese K, Clune S (2014) The influence of packaging attributes on consumer behaviour in food-packaging life cycle assessment studies—A neglected topic. J Clean Prod 73:100–108
Williams H, Wikström F (2011) Environmental impact of packaging and food losses in a life cycle perspective: a comparative analysis of five food items. J Clean Prod 19:43–48
Williams H, Wikström F, Löfgren M (2008) A life cycle perspective on environmental effects of customer focused packaging development. J Clean Prod 16(7):853–859
Williams H, Wikström F, Otterbring T, Löfgren M, Gustafsson A (2012) Reasons for household food waste with special attention to packaging. J Clean Prod 24:141–148
Xu G, Liu X, Lin Y, He G, Wang W, Xiong W, Luo H, Liu Z, Zhao J (2015a) Thermal hydrolysis of poly(L-lactic acid) films and cytotoxicity of water-soluble degradation products. J Appl Polym Sci 132(25):42064
Xu Z, Sun D-W, Zeng X-A, Liu D, Pu H (2015b) Research developments in methods to reduce the carbon footprint of the food system: a review. Crit Rev Food Sci Nutr 55(9):1270–1286
Xue X, Landis AE (2010) Eutrophication potential of food consumption patterns. Environ Sci Technol 44(16):6450–6456
Yano J, Hirai Y, Sakai S-I, Tsubota J (2014) Greenhouse gas emissions from the treatment of household plastic containers and packaging: replacement with biomass-based materials. Waste Manage Res 32(4):304–316
Zampori L, Dotelli G (2014) Design of a sustainable packaging in the food sector by applying LCA. Int J Life Cycle Assess 19(1):206–217
Zhang H, Wen Z-G (2014) The consumption and recycling collection system of PET bottles: a case study of Beijing, China. Waste Manage 34(6):987–998
Zhang SF, Zhang LL, Luo K, Sun ZX, Mei XX (2014) Separation properties of aluminium-plastic laminates in post-consumer Tetra Pak with mixed organic solvent. Waste Manage Res 32(4):317–322

Sustainable Design of Packaging Materials

Sara González-García, Esther Sanye-Mengual, Pere Llorach-Masana, Gumersindo Feijoo, Xavier Gabarrell, Joan Rieradevall and Maria Teresa Moreira

Abstract The development and production of products in a more sustainable way has received special attention in recent years. In particular, packaging products range from single materials with simple designs as well as complex ones that include different materials (cardboard, woody boards, paper, plastics, etc.). A comprehensive assessment of the environmental impacts of a product's life cycle comprises functions from the extraction of raw materials to waste management and disposal (i.e., the life cycle-assessment perspective). Thus, the knowledge of the environmental impacts of packaging products used in a specific production sector is a factor of major importance not only with the aim of improving the environmental performance of products and/or processes but also to fulfill the requirements of the ecological/green products market. One of the most valid tools to assess and reduce the inherent environmental burdens associated with products is ecodesign or Design for the Environment (DfE). This methodology consists of applying environmental criteria to the development of a product and implies a change of how we regard that product. The assessment of environmental

S. González-García · G. Feijoo · M.T. Moreira (✉)
Department of Chemical Engineering, Institute of Technology,
University of Santiago de Compostela, 15782 Santiago de Compostela, Spain
e-mail: maite.moreira@usc.es

E. Sanye-Mengual · P. Llorach-Masana · X. Gabarrell · J. Rieradevall
Sostenipra (ICTA-IRTA-Inedit) (SGR1412)-Institut de Ciència i Tecnologia Ambientals (ICTA),
Universitat Autònoma de Barcelona (UAB), Bellaterra, Spain

P. Llorach-Masana
ELISAVA Barcelona School of Design and Engineering, Barcelona, Spain

J. Rieradevall
Department of Chemical Engineering (XRB), UAB, Bellaterra, Spain

S.S. Muthu (ed.), *Environmental Footprints of Packaging*,
Environmental Footprints and Eco-design of Products and Processes,
DOI 10.1007/978-981-287-913-4_2

improvement of the product's entire life cycle is also considered for a comprehensive analysis. To demonstrate the application of DfE in the ecodesign of packaging products, a wooden storage box was assessed. Different types of materials, such as timber, plywood, engineered woods, plastics, brads, hoods, and/or staples, can be considered in the manufacture process. This type of box is often used for packaging when mechanical resistance is required for heavy loads, long-term warehousing, or adequate rigidity. Moreover, when such a box is used in the food sector, its production chain must include fitosanitary thermal treatment. According to the assessment by means of DfE methodology, the relevance of the raw materials chosen, as well as their origin, can greatly influence the associated environmental burdens, which can also be confirmed quantitatively by LCA. Thus, a correct methodological adaptation of the concept of "eco-briefing" as a tool for communication among environmental technicians and designers, includes the simplification of the analytical tool used and the application of the life cycle-assessment methodology, which facilitates the environmental analysis, are required to obtain new formats of packaging materials designed within a sustainable perspective.

Keywords Design for environment · Ecodesign · Environmental performance · Life cycle assessment · Materials selection

1 Introduction

Environmental issues, such as climate change and fossil fuels depletion, have led to a society hat increasingly aware of environmental preservation (Ribeiro et al. 2013). One of the major aspects in the process of product development is the one related to materials selection, which is not only associated with products manufacture but also with packaging (González-García et al. 2011a; Sanyé et al. 2012; Peças et al. 2013). Therefore, the growing concern about products being manufactured in a sustainable manner involves paying special attention to packaging materials. Different investigators have reported the outstanding contribution from environmentally friendly packaging of a wide range of products in the context of global environmental impact (Koreneos et al. 2005; Meyhoff Fry and Edwards 2011; González-García et al. 2011a; Sanyé et al. 2012). According to these studies, a good packaging design could contribute to decreases in the environmental impact of a product as well as lower production costs (Ribeiro et al. 2008).

Packaging products have a strong presence in markets as well because they have turned into essential elements in the life cycle of other products. In fact, packaging has the function of protecting and maintaining products during the distribution and retail processes all the way to the final user (Sanyé-Mengual et al. 2014a). Specifically in the food sector, advances in food packaging play a major role in keeping the food supply safe (Marsh and Bugusu 2007; Meyhoff Fry and Edwards 2011). Packaging technology must balance food protection with other issues including energy and material costs, social and environmental awareness, and compliance with regulations on the disposal of municipal solid waste (Jungbluth et al. 2000; Marsh and Bugusu 2007; Madival et al. 2009).

Multiple examples exist of reporting the environmental impacts of packaging materials in the food sector (Spitzley et al. 1997; Koreneos et al. 2005; Siracusa et al. 2008; Meyhoff Fry and Edwards 2011; Antón et al. 2008; González-García et al. 2013a, b) including a remarkable case study on a sparkling drink[1] regarding the introduction of both new packaging designs and recycling concept (Sanyé-Mengual et al. 2014a). Thus, packaging has evolved into a new integral part of the product where design and marketing play an imperative task. The environmental burdens of products are increased due to not only the amount and type of packaging materials (Jungbluth et al. 2000) but also the packaging-material management approach (Ross and Evans 2003; Büsser and Jungbluth 2009; Sanyé-Mengual et al. 2014a). Therefore, proper management of these packaging wastes is also important in terms of environmental consequences (recycling, reuse, valorization, landfilling, etc.). To comply with the current European legislation on packaging and packaging waste (European Council 1994, 1997, 2004, 2005, 2009), packaging producers must take all possible measures to reduce the environmental impact of packaging products while retaining the functions that existed prior to the admission of the product in the market.

Although numerous studies have quantified the environmental consequences derived from packaging materials, the influence of the packaging during the full life cycle of products is reasonably different depending on the product considered (Jungbluth et al. 2000). Particular attention is being paid to utilizing alternative raw materials specifically for polymers (Siracusa et al. 2009). So far, petroleum-based polymers have been used as packaging materials due to their large availability at relatively low costs as well as good insulating and mechanical properties (Siracusa et al. 2009). Substitutes for plastic packaging (such as steel, aluminum, glass, cardboard, packaging paper) vary depending on the market sector and packaging application. In this sense, cork and rubber are alternatives in the caps and closures category (Franklin Associates 2014). However, plastic packaging also presents disadvantages because they are not completely recyclable and/or biodegradable. In this sense, research is being focused on the development of biodegradable polymers and bioplastics made from renewable raw materials (Siracusa et al. 2009; Moralejo-Gárate et al. 2013; European Bioplastics 2015).

This chapter focuses on the process of applying environmentally friendly strategies in the design of packaging products. First, environmental strategies that can be included in the life cycle of packaging products are proposed. Second, the methodology to improve the design of packaging by combining design for environment (DfE) and life cycle assessment (LCA) is described. Finally, a case study of a wooden storage box is assessed.

[1]http://www.carbontrust.com/media/5888/cts287-coca-cola.pdf.

2 Integration of Environmental Aspects into Packaging Design

In the framework of design for environment (DfE), a large number of ecodesign strategies have been proposed to improve the environmental performance of products. All of them are commonly grouped according the life-cycle stage they affect (Crul and Diehl 2006; van Hemel 1998). This section aims to select and collect those ecodesign strategies than can be applied to the packaging sector where packaging is analyzed as a single product rather than as part of a life-cycle stage. These strategies may be used as a guidance source for designers and policy makers when applying ecodesign to packaging products.

Tables 1 and 2 show the recommended ecodesign strategies for the packaging sector by life-cycle stage. The list of environmental strategies in DfE provided by Sanyé-Mengual et al. (2014c) was combined with a new set of specific strategies for packaging products. The tables include strategies for the following life-cycle stages: concept, materials, production distribution, and end-of-life. Strategies for the use stage were omitted because they do not apply for packaging products or are covered in other stages such as the concept stage.

The concept stage (Table 1) usually has a great potential to reduce the environmental impact of products (van Hemel 1998). However, applying strategies such as dematerialization may sometimes require redesigning a product and generating new concepts. Packaging products have already been optimized during previous years (Bovea and Gallardo 2006; Sanyé-Mengual et al. 2014a). Consequently, achieving strong modifications for dematerialization in packaging products could result in a difficult task for companies. Nevertheless, work can be done with little investment to increase the environmental information included in packaging products.

According to Table 1, strategies for packaging materials have great potential to reduce the environmental impact of these products. In this stage, three specific strategies were added for packaging products: (1) the use of natural printing inks; (2) the avoidance of adhesives or use of natural ones; and (3) the avoidance vinyls and stickers. These strategies are oriented to reduce the environmental impact when integrating packaging into communications support (e.g., the brand or products' properties). Communication within packaging (e.g., use of stickers) could lead to difficulties in separating materials for recycling. Moreover, the use of synthetic inks would increase the environmental impact of packaging at the end-of-life stage.

Table 2 displays the common environmental strategies to improve the production stage for all types of products. These strategies apply for many different production processes. However, their potential environmental benefits are dependent on the best technologies available. The main objectives in this stage are to reduce resource and energy consumption as well as waste generation.

For the distribution stage, strategies are oriented to increase the efficiency of the transportation process and, consequently, to optimize the volume and weight

Table 1 Environmental strategies for the concept and material stages of a packaging product by life-cycle stage and benefit

Strategies	Reduced resources consumption	Reduced environmental impact	Reduced energy consumption	Enhanced recycling/reusing	Decoupling from non-renewable resources	Increased lifespan	Market differentiation	Improved user behaviour
Concept								
Dematerialization	•	•						
Multifunctionality						•		
Environmental information (e.g. carbon footprint)							•	•
Demand of suppliers' environmental information		•						
Ensure packaging durability if reused		•						
Shared use of packaging (standard packaging)	•	•		•				
Materials								
Dematerialization	•	•						
Monomaterial				•				
Recyclable materials				•				
Renewable/natural resources		•			•			
Low-impact materials		•						
Local resources		•					•	
Reused components	•	•						
Use of natural printing inks		•						
Avoid adhesives or use natural ones		•						
Avoid vinyls and stickers	•	•		•				

Table 2 Environmental strategies for the production, distribution, and end-of-life stages of a packaging product by life-cycle stage and benefit

Strategies	Reduced resources consumption	Reduced environmental impact	Reduced energy consumption	Enhanced recycling/ reusing	Decoupling from non-renewable resources	Increased lifespan	Market differentiation	Improved user behaviour
Production								
Internal recycling (closed-loop)	•							
Optimize production process	•	•	•					
Choose cleaner production processes	•		•					
Use of low-impact energy sources		•						
Promote renewable energy sources					•			
Local production		•	•					
Distribution								
Optimization of product weight	•		•					
Optimization of packaging volume		•	•					
Local distribution		•	•					
Bio-fuels transportation		•	•					
Efficient transportation		•	•					
End of life								
Feasibility of components separation				•				
Recyclability				•				
Materials identification				•				
Reusability				•		•		
Biodegradability				•				
Communication-to-user (waste management)		•						•

of the packaging product or to use more energy-efficient transportation vehicles. However, in this case, strategies such as optimizing the volume and weight of the packaging are dependent of the product being packaged. Consequently, these strategies may be developed accordingly with the requirements and properties (e.g., dimensions) of the particular product.

Finally, environmental strategies to improve the end-of-life packaging of products are very similar to all type of products. These are basically oriented to reduce resource consumption by enhancing the reusability of elements or by promoting its recycling. Increasing the use of biodegradable materials, or communicating to the user the optimal ways to manage this product as a waste, aims to reduce the environmental burdens of this stage.

As mentioned previously, the stage with more specific strategies for packaging products is the materials stage due to the requirements of packaging to communicate information. Using inappropriate technologies for adding information on packaging products could result in a significant environmental impact. For the other stages, the strategies mentioned are in common use for different type of products such as furniture or textiles.

3 Design for Environment Methodology

3.1 Introduction

Although LCA methodology is a suitable and valuable tool to assess the environmental impact of materials during their life cycle (Baumann and Tillman 2004), it can also be combined with environmental tools to analyze and reduce the environmental burdens associated with products.

Ecodesign or Design for the Environment (DfE) is receiving special attention as a potential instrument in product-development strategies. Product design is one of the most important production strategies toward global sustainability due to the fact that all products available in markets are the result of a product-development process (Ramani et al. 2010). DfE integrates multifaceted aspects of both design and environmental considerations. It takes into account that the definition of sustainable solutions for products must be based on the minimization of negative consequences in the context of economic, environmental, and social perspectives (Charter and Tischner 2001).

This methodology is composed of applying environmental criteria to the development and design of a product (Ramani et al. 2010). So although many other definitions exist, DfE is considered as the design of and for a sustainable development context (Karlsson and Luttropp 2006). This change in the design process is translated into (1) a reduction of environmental emissions and (2) the improvement of the environmental profile of products throughout the entire life cycle taking all the involved steps into consideration (McDonough et al. 2003; Zust and Wirnmer 2004).

Consequently, LCA and DfE constitute a good relationship because LCA provides the structure for analyzing the environmental impacts associated to a product and DfE can perform the practical application of the assessment (Ramani et al. 2010).

3.2 Stages of Design for Environment

DfE refers to the methodical integration of environmental factors into product design and development, thus playing a crucial role in the development of an integrated product policy (Tukker et al. 2000; Sanyé-Mengual et al. 2014a).

Certain environmental objectives must be set for a proper conceptual development, based on which, and by means of a critical review by a panel of expert participants, the process of ecodesign is initiated with consideration of all the stages of the life cycle (Smith and Wyatt 2006). Thus, a proper and fluid communication between environmental experts and designers is mandatory. Figure 1 displays the different steps to fulfil in an ecodesign strategy.

Step 1. Establishment of the multidisciplinary ecodesign team

An important aspect that must be considered is the creation of a multidisciplinary team to cover the different fields of knowledge involved not only in the design and environment but also in the manufacture process. Commonly the ecodesign team is constituted by designers, engineers, environmental scientists, chemists, and experts in the field of the industrial product under study.

Step 2. Description of variables that define the product to ecodesign

This phase of ecodesign strategy requires special attention because both the type and number of variables to be analyzed depend on the product selected for the assessment. Thus, selection criteria must be established to prioritize potential variables that could also be applied to similar products. Aspects related to the product (and sector), such as implementation and complexity degree, representative materials, as well as market demands, are compulsory.

Step 3. LCA of the selected product

This phase of DfE is based on the environmental assessment of the product chosen for ecodesign by means of LCA methodology. Therefore, not only is the environmental profile derived from the life cycle of the product determined, the significant environmental factors (also known as environmental "hot spots") are also identified. This step is the starting point for the eco-briefing.

Step 4. Establishment of eco-briefing and ecodesign strategies

Eco-briefing involves the environmental aims that should be considered in the development of ecodesign strategies and is the procedure to communicate the most suitable strategies. Consequently, the environmental goals established to be achieved by means of ecodesign must be carefully indicated. Ecodesign strategies are the alternatives that eco-briefing addresses with the aim of improving the

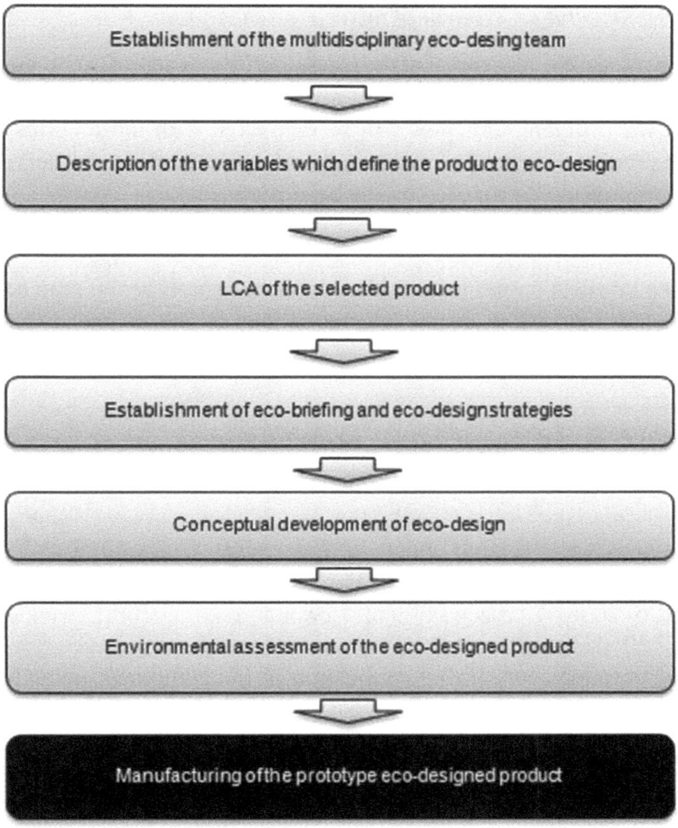

Fig. 1 Steps in the DfE methodology for the ecodesign of a general product

current environmental performance of the selected product (Bhamra 2004; Ferrao and Amaral 2006) by analysing not only technological but also social and financial aspects (i.e., the sustainable perspective). Key life-cycle stages under consideration in eco-briefing are product conceptualization, materials used, production process, distribution, maintenance, and end-of-life management.

Step 5. Conceptual development of ecodesign

Once the ecodesign strategies are defined, a conceptual line to be followed is defined. Special attention should be paid to the key life-cycle stages that have received more attention in the eco-briefing step (i.e., higher punctuations) as well as to the most viable strategies (in terms of technological, financial, and social issues) taking into account feedback from the ecodesign team. These strategies with higher viability to be implemented are the ones to be assessed as well as classified as quantitative and qualitative alternatives. Thus, this step requires a continuous relationship between the team partners in order to analyse in situ the development of the ecodesign.

Step 6. Environmental assessment of the ecodesigned product

This step involves the environmental assessment by means of LCA methodology of the proposed viable quantitative ecodesign strategies. Afterward, environmental profiles for the different strategies will be compared with those corresponding to the current product. The aim of this comparison is to analyze the degree of environmental improvement proposed by the ecodesign team.

Step 7. Manufacturing of the prototype ecodesigned product

The last step consists on the manufacturing of the prototype, i.e., the ecodesigned product, according to the strategies selected in Step 6.

3.3 Products Ecodesigned by a Combination of LCA and DfE Methodologies

Multiple studies are available about the procedure of ecodesign and its interest in the development of integrated product policy (Bovea and Vidal 2004; Bovea and Gallardo 2006; Kurczewski and Lewandowska 2010; Lewandowska and Kurczewski 2010; Tukker et al. 2000).

Practical examples concerning application of the combined methodologies for ecodesign can be found in very different industrial sectors: the automobile sector (Ruhland et al. 2004; Finkbeiner et al. 2006; Muñoz et al. 2006), the leather tanning industry (Rivela et al. 2004), the packaging sector (Bovea and Gallardo 2006; Sanyé-Mengual et al. 2014a), cutlery (Sanyé-Mengual et al. 2014b), clothing (Sanyé-Mengual et al. 2014b), electronic devices (Nedermark 1998; Mathieux et al. 2001; Aoe 2007; Gazulla et al. 2007; Unger et al. 2008), lighting (Gottberg et al. 2006; Casamayor and Su 2013), printing (Tischner and Nickel 2003), and waste management (Todd et al. 2003). Special attention has been paid to wood-based materials. Numerous studies are available in the literature where ecodesign strategies have been applied to wood-based products especially due to the interest in the procurement of wooden goods produced in a sustainable manner as well as in giving solutions to the wood-production sector. Examples include wood boards (Bovea and Vidal 2004), woody surface and edge coverings (Bovea and Vidal 2004), modular playgrounds (González-García et al. 2012a), child furniture sets (González-García et al. 2012b), goods containers (González-García et al. 2011a), kitchen cabinets, office tables, and ventilated walls and headboards (González-García et al. 2011b, 2012c). According to all these studies, the process of integrating the environmental aspects into product development is only effective if it leads to an improved product with fewer environmental impacts and if communicating maintenance procedures to consumers form part of the ecodesign process (Sanyé-Mengual et al. 2014b).

4 Case Study: Storage Wood Box

4.1 Description of the Case Study and Product Under Assessment

As mentioned previously, changes in the design process can promote reductions of environmental impacts. Thus, the interdisciplinary team involved in design for the environment plays a major role in the improvement not only in the ecodesigned product but also in the product-production stages.

The ecodesign of a wood product, such as a storage box, was proposed for assessment. The interest in this product is justified because wood boxes are extensively used not only for storage products but also for transport activities and are present in multiple different sectors and activities. Thus, this section of the chapter reports the methodology used to perform ecodesign of the wood box taking into account its manufacturing process as well as the eco-briefing strategies over all of the key life-cycle stages. Moreover, the environmental impacts derived from woody boxes production are determined using LCA methodology.

To do so, representative primary data were procured directly from a Spanish company located in Galicia (Northwest Spain), that is a Spanish leader in terms of wood-based boards and wood-derived products such as boxes. Although different types of wood boxes are produced, we paid attention to those destined to be in the wine sector. The box considered for assessment is typically used for the storage of three standard wine bottles (750 mL) and presents the following dimensions: 350 × 260 × 103 mm with an average weight of 1.35 kg (González-García et al. 2011a). Because the production process could be considered representative for the manufacture of other wood boxes with different uses and dimensions, two functional units were considered for assessment. Therefore, the ecodesign study is reported in terms of one woody box with the dimensions aforementioned. In addition, we considered 1 kg of wood box as alternative functional unit in order to report the environmental results corresponding to the production system (Fig. 2).

The specific box considered for assessment mainly consists of MDF (medium density fiberboard) and solid timber joined with metal pieces such as brads, hoops, and staples. The wood-box production system was divided into three steps taking into account the primary activities carried out in the factory: the manufacturing step (including assembling, painting, and packaging processes), the cogeneration step in order to produce the energy requirements, and the distribution step to clients. Secondary activities related to the production and transportation of different inputs to the system, e.g., chemicals, boards, metal pieces, or ancillary packaging materials, were also taken into account and computed within the system boundaries (Fig. 2). According to the system boundaries depicted in Fig. 2, further activities related to woody-box use, maintenance, and final management were excluded from the assessment due to the lack of real and valuable information and inventory data. Moreover, these further activities are beyond the premises of the woody factory under assessment. For that reason, a cradle-to-gate perspective was considered in this case study.

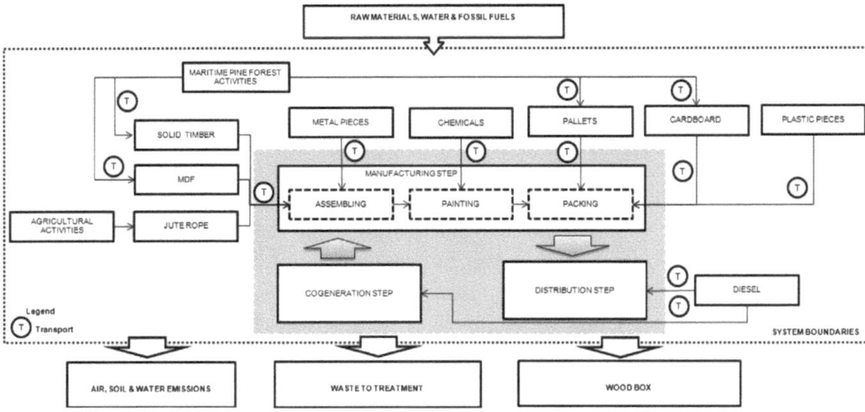

Fig. 2 System boundaries and processes included within the analysis

The Life Cycle Inventory (LCI) data for the foreground system, which includes all the activities carried out in the factory considered for assessment, were collected by means of surveys and interviews with workers. Whenever possible and feasible, typical process-specific data of a period of 1 year were collected. Secondary data corresponding to the production of different inputs were taken from databases (González-García et al. 2011a). Thus, inventory data corresponding to the production of metal pieces (staples, brads, and hoops) were taken from the IDEMAT database (2001). Inventory data for the remaining background processes—such as these corresponding to the production of plastic pieces (hoops and film), the production of the alkyd paint used in the painting process, the production of the jute rope for the handle, the production of the solid timber, and the production of wood pallets—were taken from the Ecoinvent database®.[2]

Concerning the production of the MDF boards, primary data from the inventory stage were taken from a previous study (Rivela et al. 2007) where three factories, considered representative of the "state of art," were evaluated. Finally, regarding forest operations for the different woody inputs (MDF, solid timber, pallets, and cardboard), inventory data were taken from González-García et al. (2013c).

When setting LCA boundaries, it must be decided whether the production and maintenance of capital goods are included within the system boundaries. In this study, they were excluded from the system boundaries because it was assumed to be comparable with that of plants producing functionally similar materials (Jungmeier et al. 2002). Allocation, an important issue in LCA studies, consists of assigning the input and/or output flows of a process to the product system under study. It is required for multifunctional processes, and the selection of an allocation approach can have a strong effect on the results. A characteristic of this woody industry is the concurrent production of very different woody products

[2]http://www.ecoinvent.org/database/.

such as panels, boxes, and papers. Thus, an allocation procedure was considered to allocate the environmental burdens between the different coproducts. There are several allocation methods (mass, economic, etc.), each of which have advantages and disadvantages. Moreover, the choice of allocation procedure depends on the limitations of the study. In this case study, mass allocation was assumed taking into account the annual production of the different coproducts. Economic allocation was not considered because it was not possible to find market prices for all of the products produced in the mill.

4.2 Environmental Perspective of the Woody Box Under Analysis

An attributional LCA for the woody box production was carried out according to the CML 2 baseline 2000 V2.1 method to quantify the environmental impact (Guinée et al. 2001). This method results in the definition of an environmental profile for the assessed product/process/service by quantifying the environmental effects on different categories, whereas only indirect or intermediate effects on humans can be assessed. The impact categories analysed in this study were as follows: abiotic depletion (ADP), acidification (AP), eutrophication (EP), global warming (GWP), ozone layer depletion (ODP), and photochemical oxidant formation (POP). The software SimaPro 8.0.2 was used to implement and process the inventory data (PRé Consultants 2014). The results for the characterisation step are shown in Table 3 per both functional units (one woody box and 1 kg of woody box).

Figure 3 displays the relative contributions from the woody box production steps in the different impact categories considered.

According to the results shown in Fig. 3, the manufacturing step is the most important stage considered throughout the production chain, with contributions ranging from 60 to 90 % depending on the category, followed by the cogeneration stage (ratios from 5 to 30 %).

The remarkable contributions in all of the categories considered are due to the fact that this step includes three relevant processes (assembling, painting, and packaging), which involve the requirements of material inputs such as

Table 3 Characterisation results per impact categories considered under evaluation

Impact category	Unit	1 woody box	1 kg woody box
Abiotic depletion (ADP)	kg Sb_{eq}	5.18×10^{-3}	3.84×10^{-3}
Acidification (AP)	kg $SO_{2\,eq}$	7.56×10^{-3}	5.60×10^{-3}
Eutrophication (EP)	kg $PO_4^{-3}{}_{eq}$	8.16×10^{-4}	6.05×10^{-4}
Global warming (GWP)	kg $CO_{2\,eq}$	6.44×10^{-1}	4.77×10^{-1}
Ozone layer depletion (ODP)	kg CFC-11 $_{eq}$	8.31×10^{-2}	6.15×10^{-2}
Photochemical oxidation (POP)	kg $C_2H_{2\,eq}$	3.45×10^{-4}	2.56×10^{-4}

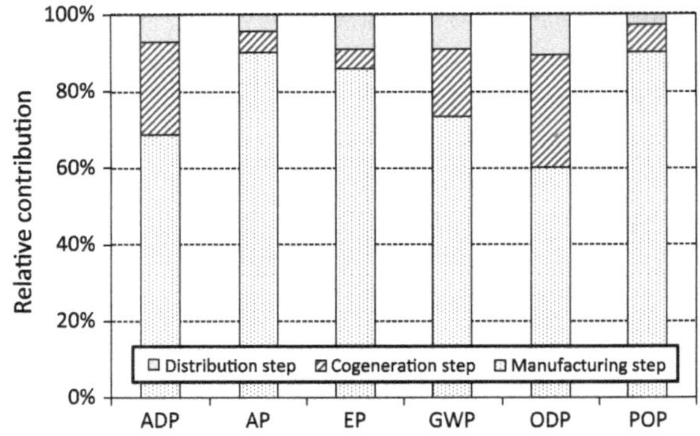

Fig. 3 Relative contributions from the woody box production steps within the system boundaries in the different impact categories considered

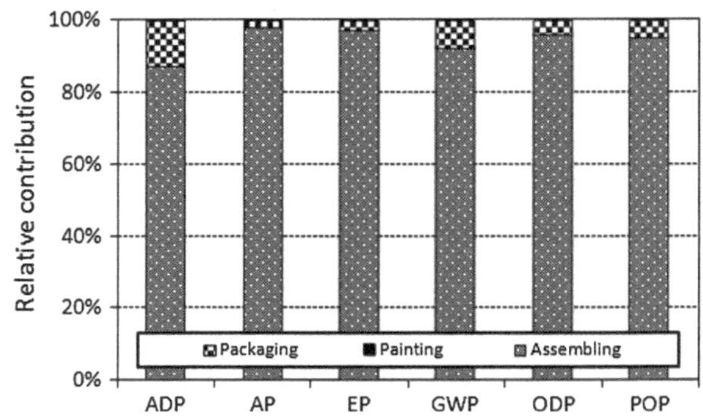

Fig. 4 Relative contributions per processes involved in the woody-box production chain within the manufacturing step

MDF boards and metal pieces, the background production activities of which are energy- and material-intensive. Therefore, a detailed assessment was proposed to analyze in detail the contributions derived from these foreground activities taking into account the corresponding background processes. Figure 4 shows the distribution of impacts (per impact category) between the foreground processes carried out in the factory. According to that figure, the assembling process is responsible for 94 % of environmental impacts derived from the manufacturing step with contributions from the painting process being almost negligible.

The assembling process is the activity during which the woody box is manufactured using MDF boards and solid pine timber as main raw materials. The construction pieces are joined with metal pieces, such as brads, hoops, and staples as well as jute rope, which is used for the handle. All of these structural materials involve background activities regarding their production and transportation up to the woody-box factory gate. Figure 5 shows the distribution of environmental impacts per factor involved in the assembling process.

According to Fig. 5, the production and distribution of the jute rope used for the handle is the main environmental hot spot in all the categories considered for assessment followed by activities related to the production of MDF boards. Thus, improvement strategies in the ecodesign should be focused in these materials used in the woody-box structure.

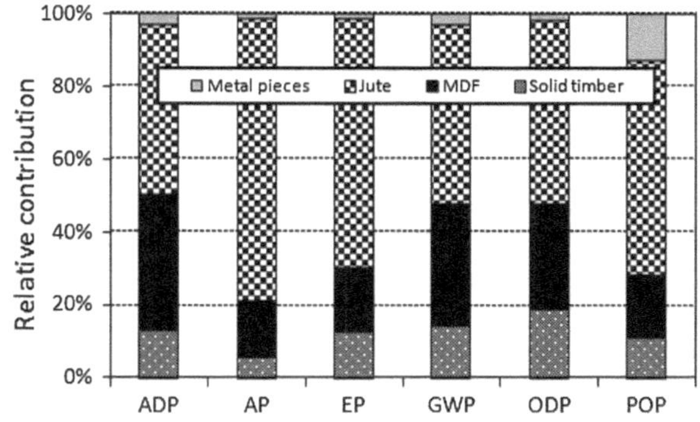

Fig. 5 Distribution of impacts per background processes involved in the assembling process

Regarding the cogeneration step (Fig. 3), this is the second most important step in terms of environmental impacts. In the factory, all of the energy requirements are produced on-site by means of the combustion of fossil fuel with low sulphur content. Production of the fossil fuel, from transport up to the factory, as well as derived combustion emissions were computed in the cogeneration step. The use of an alternative renewable source to produce the energy requirements could be an interesting improvement alternative to take into account.

4.3 Ecodesign of the Woody Box

As was defined in Sect. 3, eco-briefing is the adaptation of a method that assists the communication of environmental factors among environmental experts and designers using basic information about the product to be designed and defining the product with the environmental objectives to be achieved. The sequence of stages proposed in Fig. 1 must be followed in a DfE study. The multidisciplinary team is comprised of environmental technicians as well as designers and other technicians from the factory involved in the production chain. Five key cycle stages were proposed for the eco-briefing: concept (C), materials (M), production (P), distribution (D), and end-of-life (E). The results from the eco-briefing are summarised in Table 4.

Thus, different strategies were proposed to obtain a woody box with a low environmental impact taking into account the results from the eco-briefing. These

Table 4 Environmental hot spots and life-cycle stages considering in the eco-briefing

Environmental hot spots	Key life-cycle stages				
	C	M	P	D	E
Functionality	■	□	□	□	□
High energy and water consumption	□	□	■	□	□
High impact vehicles	□	□	□	■	□
Low optimization of transport volume	■	□	□	■	□

strategies were evaluated from a technological, economical, and social perspective. However, only the most viable strategies for the factory will be discussed below and were considered for the ecodesign.

5 Discussion of Ecodesigned Alternatives and Environmental Profiles

5.1 Alternative Materials for the Structure and Handle of the Box

As stated in the environmental analysis of the woody box, the assembling step produces the greatest environmental impacts with the MDF and the jute rope being the main responsible factors of these results. The MDF represents approximately 22 % of the total weight, thus ranking as the second most important material in terms of weight (72 % of the total weight is the solid pine timber, which is not considered to be an environmental *hot spot*). The production of this material involves large amounts of energy requirements as well as chemicals such as adhesives (Rivela et al. 2007).

According to the factory workers, alternative materials, such as pine plywood or even solid pine timber, could be used as a substitute for MDF without changing the woody-box properties and characteristics. The use of these alternative materials should also produce changes in the total weight of the woody box due to differences in their density (González-García et al. 2011a). Thus, the weight of the current box (1.35 kg) should be reduced by approximately 10 % (approximately 1.46 kg) if plywood is used as a potential structural material (approximately 1.2 kg) or increased by 8 % (approximately 1.46 kg) if solid timber is used. Regardless, for the functional unit considered to display the environmental profiles (that is, per unit box or per kilogram of box), the use of solid pine timber instead of MDF (or plywood) should produce the least environmental impacts (Fig. 6). Differences were identified in the environmental behavior depending on the functional unit (Fig. 6a, b). If the results are reported per unit box (Fig. 6a), the worst environmental profile should correspond to the current box (MDF box) in all of the categories under assessment except in terms of POP, whereas the plywood box should produce a slightly greater impact. The production of the plywood box should result in minor impact reductions ranging from 0.1 to 5 % compared with the MDF box. This slight improvement of the profile should be related with the lowest amount of board required to produce the same product, which should present the lowest impact from the plywood production. The solid pine timber box should report the best environmental results in all the categories with reductions ranging from 13 % (AP and EP) to 21 % (GWP and ADP). However, although better results would be obtained, more research should be required, specifically from design and technological issues, to make the lip of the box in one piece, which at the present moment, is problematic due to the timber width.

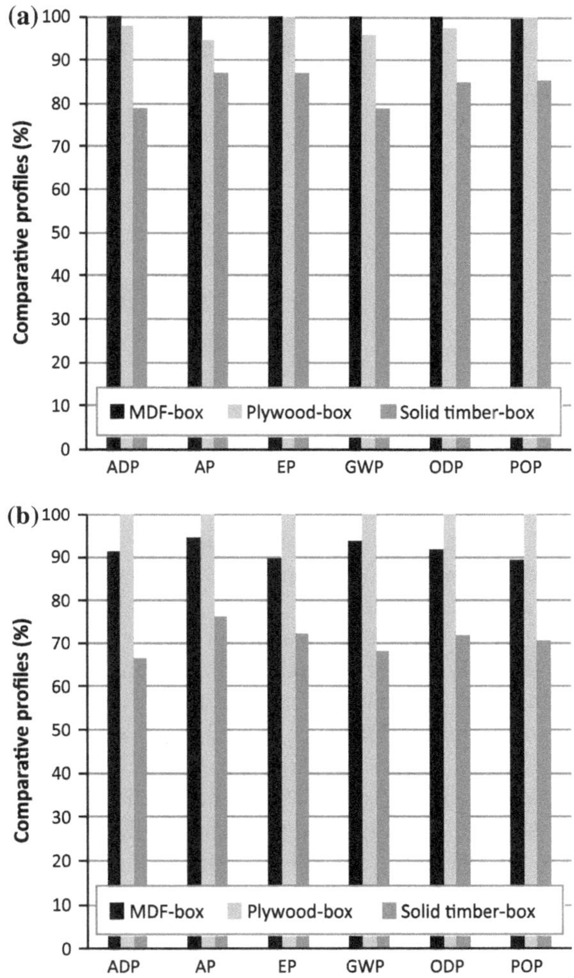

Fig. 6 Comparative environmental profiles considering alternative structural materials. **a** Profiles per unit box; **b** profiles per 1 kg of box

However, if the comparison is carried out per kilogram of woody box (Fig. 6b), the plywood box should present the worse profile in terms of all of the impacts considered, with once again the solid pine timber box having the best profile. Thus, despite reducing the total weight of the box by 10 % when substituting MDF with plywood, the impacts should increase in ratios ranging from 6 to 12 %. The highest chemical and energy requirements in the plywood production process should be the responsible actors of these "negative" results.

Another improvement action to consider for the ecodesign of the woody box should focus on the substitution of the fibres used in the handle (jute rope) by other alternative fibres with similar properties that are available on the market. The jute rope processed in the factory is transported from India, which accounts for large impacts due consumption of energy for transportation (González-García et al. 2011a). Thus, the use of national or regional fibres is expected to report

Fig. 7 Comparative environmental profiles considering alternative fibre materials

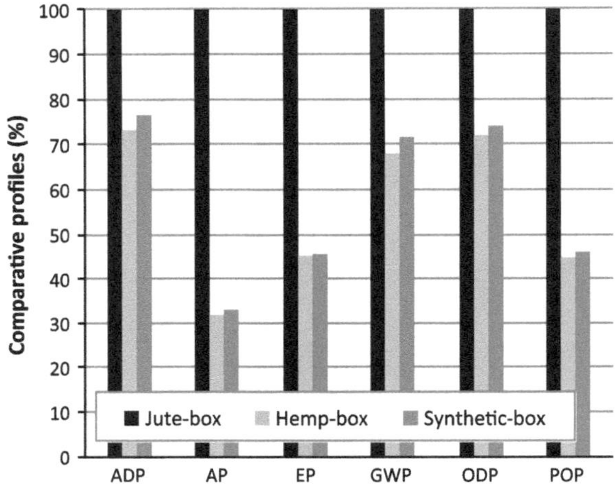

better environmental profiles compared with jute fibres. Two alternative fibres were proposed for assessment by factory workers: (1) hemp fibres, which are extensively cultivated in Catalonia (González-García et al. 2010); and (2) synthetic fibres from Madrid (González-García et al. 2011a). Comparative profiles are displayed in Fig. 7. In this case and regarding differences in the types of boards, no changes are expected for the amount of fibres required to produce the handle, so the environmental changes should be based on differences in transport distances as well as the fibre-production processes. Therefore, the same comparative profiles should be obtained regardless of the functional unit.

According to Fig. 7, the alternative fibre material considered to substitute in place of the jute rope for the handle should result in important environmental improvement specifically in terms of AP, PE, and GWP. It is important to highlight the remarkable effect from the transport activities and thus promote of the use of national fibres.

5.2 Alternative Energy Sources in the Cogeneration Step

The cogeneration step was (by far) the second most important foreground step (Fig. 3). All energy requirements (heat and electricity) are produced on-site using low-sulphur diesel fuel, such as fossil fuel, in the cogeneration unit. Thus, important contributions to impact categories, such as ADP, GWP and ODP, have previously been reported. Therefore, an alternative was proposed based on the use of a renewable energy source, such as wood chips, to promote the use of bioenergy. As expected, remarkable improvements should be achieved with the use of wood chips as fuel in the cogeneration unit because the cogeneration step is remarkable among all of the categories under assessment (Fig. 8).

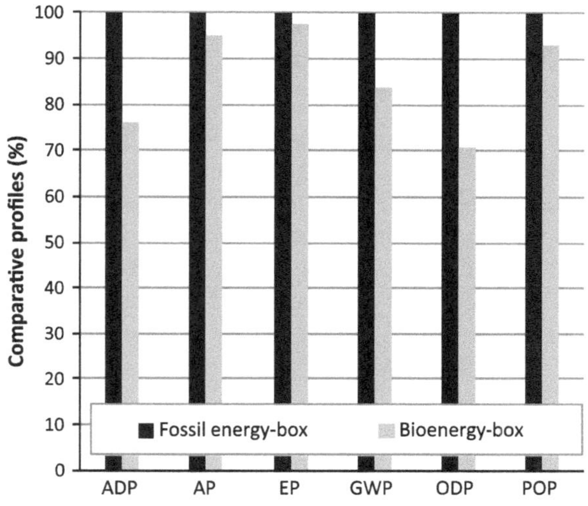

Fig. 8 Comparative environmental profiles considering alternative fuel sources

5.3 Ecodesigned Woody Box

According to the eco-briefing, ecodesign strategies should increase the functionality of the wood box. Thus, an increment in the functionality of the box should result in a longer life span and thus more intensive use of the box. The alternatives reported previously that had the best environmental profiles were also considered in the ecodesigned woody box. Thus, production should include the use of only solid pine timber as structural material, a hemp fibre-based handle, bioenergy from wood chips, and a conceptual proposal for the woody box as bird nest box for increased functionality. Environmental (as well as social) improvements are shown in Fig. 9. Thus, benefits in all of the categories under analysis should be achieved by increasing the sustainability of the woody box.

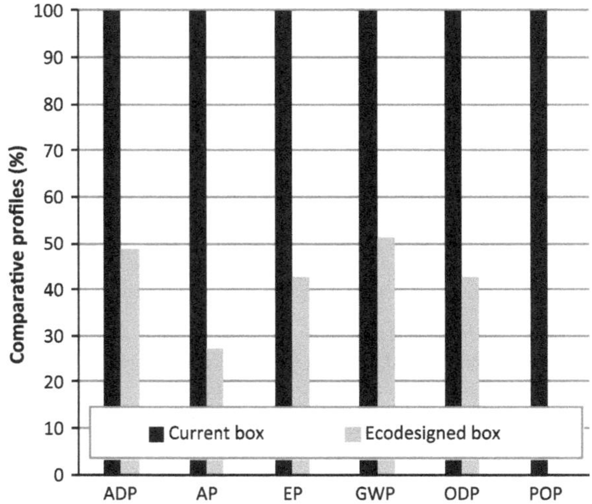

Fig. 9 Comparative environmental profiles between the current and ecodesigned woody box

6 Limitations and Recommendations on Ecodesign for Packaging Materials

The implementation of ecodesign strategies can be constrained because the two main functions of packaging must be preserved, thus becoming imperative requirements: (1) ensure the protection of packed products; and (2) guarantee a good communication of the corporate image for both product and company.

Some ecodesign strategies (e.g., improve the logistics of the product) are not strongly affected by the mentioned requirements. However, other ecodesign strategies with great potential to reduce the environmental burdens of the product (e.g., use of local, renewable, or recycled materials) might be limited by these requirements. Consequently, the environmental benefits of some of the ecodesign strategies depend on the creativity of the industrial designers who use them while ensuring the good structural and communication properties of the packaging. The wooden-box case study analyzed is a clear example of how the originality required to increase the functionality of the box helped to significantly reduce the environmental impact of the product. In some cases, a complete redesign of the packaging could also be required when applying some specific ecodesign strategies. Due to the usual simplicity of packaging products, all of these achievements could be difficult to attain.

As has been demonstrated, ecodesign strategies allow improving the environmental performance of packaging materials, thus saving energy and materials. Moreover, ecodesign combined with LCA allows introducing and developing alternatives in the production processes that can be implemented for short or long periods of time. Eco-briefing is a tool for communication among environmental technicians and designers whose results, together with environmental results, can facilitate environmental analysis.

7 Conclusions

Packaging has a large presence in the market because packages are used for the protection and distribution of products. Environmental strategies applied to this sector can positively affect the environmental burdens of the products for which packaging products are part of their life cycle. Thus, the use of ecodesign as a tool to improve packaging can result in large ecological improvements.

One of the main issues in packaging design is material selection, which determines aspects such as the recyclability or the use of renewable materials. Even more, the origin of these materials can result in large environmental burdens. Thus, ecodesign strategies may lead to a better selection of packaging materials by prioritizing the use of local raw materials, dematerialization and weight reduction, and the use of recyclable materials.

Multifunctionality has been pointed out in ecodesign as an optimal environmental strategy because then the environmental burdens can be distributed among the multiple functions provided by a product. However, in the case of packaging, ecodesign can result in a complete redesign of the product, which is not applicable to all cases. Furthermore, packaging already provides two functions by providing protection and informing consumers.

To ensure that consumers perform a suitable waste-management practice when disposing of packaging products, communication regarding the product end-of-life is essential. Furthermore, this consumer education may also accomplish the expected environmental impact of the product's entire life cycle accounted for by the designers. Graphic solutions to perform this and other communications require further studies in order to determine the best available technology in environmental terms.

The case study described in this chapter highlighted the usefulness of ecodesign. The combined method of design for environment (DfE) and life cycle assessment (LCA) resulted in a useful tool for designers. The selection of environmental strategies and their quantitative potential were essential in the decision-making process.

Acknowledgements The authors (S. González-García, G. Feijoo and M.T. Moreira) belong to the Galician Competitive Research Group GRC 2013-032, programme co-funded by FEDER. Dr. Sara González-García would like to express her gratitude to the Spanish Ministry of Economy and Competitivity (Grant reference JCI-2012-11898) for financial support.

References

Antón A, de León WE, Raya V et al (2008) Life cycle assessment applied to tomato production in the Canary Islands. In: The 6th international conference on LCA in the agri-food sector. Zurich, 12–24 Nov 2008

Aoe T (2007) Eco-efficiency and ecodesign in electrical and electronic products. J Clean Prod 15(5):1406–1414

Baumann H, Tillman AM (2004) The Hitch Hilker's guide to LCA. An orientation in life cycle assessment methodology and application. Studentlitteratur, Lund. ISBN 9144023642

Bhamra TA (2004) Eco-design: the search for development new strategies in product. Proceedings of the Institution of Mechanical Engineers Part B. Int J Eng Science 218(5): 557–569

Bovea MD, Vidal R (2004) Materials selection for sustainable product design: a case study of wood based furniture eco-design. Mater Des 25:111–116

Bovea MD, Gallardo A (2006) The influence of impact assessment methods on materials selection for eco-design. Mater Des 27(3):209–215

Büsser S, Jungbluth N (2009) The role of flexible packaging in the life cycle of coffee and butter. Int J Life Cycle Ass 14(Suppl 1):S80–S90

Casamayor JL, Su S (2013) Integration of eco-design tools into the development of eco-lighting products. J Clean Prod 47:32–42

Charter M, Tischner U (2001) Sustainable solutions: developing products and services for the future (Sheffield. Greenleaf Publishing, UK

Crul MRM, Diehl JC (2006) Design for sustainability. TU DELF & UNEP, A practical approach for Developing economies

European Bioplastics (2015) http://en.european-bioplastics.org/
European Council (1994) Directive 94/62/EC of the European Parliament and of the Council of 20 Dec 1994 on packaging and packaging waste
European Council (1997) Decision 97/129/EC of the European Parliament and of the Council of 28 Jan 1997 on the identification system for packaging materials
European Council (2004) Directive 2004/12/EC of the European Parliament and of the Council of 11 Feb 2004 amending Directive 94/62/EC on packaging and packaging waste—statement by the Council, the Commission and the European Parliament
European Council (2005) Directive 2005/20/EC of the European Parliament and of the Council of 9 Mar 2005 amending the Directive 94/62/EC on packaging and packaging waste
European Council (2009) Regulation (EC) No 219/2009 of the European Parliament and of the Council of 11 Mar 2009 adapting a number of instruments subject to the procedure referred to in Article 251 of the Treaty to Council Decision 1999/468/EC with regard to the regulatory procedure with scrutiny—adaptation to the regulatory procedure with scrutiny—Part Two
Ferrao P, Amaral J (2006) Design for recycling in the automobile industry: new approaches and new tools. J Eng Design 5:447–462
Finkbeiner M, Hoffmann R, Ruhland K et al (2006) Application of life cycle assessment for the environmental certificate of the Mercedes-Benz S-Class. Int J Life Cycle Ass 11:240–246
Franklin Associates (2014) Impact of plastics packaging on life cycle energy consumption & greenhouse gas emissions in the United States and Canada. Substitution Analysis. Franklin Associates, A Division of Eastern Research Group (ERG). CLIENTS\ACC\KC142527. 143 p
Gazulla C, Fullana P, Bala A (2007) ECOTOYS—ecodesign of electric and electronic toys. SETAC Europe 17th annual meeting. Porto (Portugal)
González-García S, Hospido A, Moreira MT et al (2010) Life cycle assessment of raw materials for non-wood pulp mills: Hemp and flax. Resour Conserv Recy 54:923–930
González-García S, Silva FJ, Moreira MT et al (2011a) Combined application of LCA and eco-design for the sustainable production of wood boxes for wine bottles storage. Int J Life Cycle Ass 16(3):224–237
González-García S, Gasol CM, García Lozano R et al (2011b) Assessing the global warming potential of wooden products from the furniture sector to improve their ecodesign. Sci Tot Environ 410:16–25
González-García S, Garcia Lozano R, Buyo P et al (2012a) Eco-innovation of a wooden based modular social playground: Application of LCA and DfE methodologies. J Clean Prod 27:21–31
González-García S, Garcia Lozano R, Moreira MT et al (2012b) Eco-innovation of a wooden childhood furniture set: An example of environmental solutions in the wood sector. Sci Tot Environ 426:318–326
González-García S, Garcia Lozano R, Estévez J et al (2012c) Environmental assessment and improvement alternatives of a ventilated wooden wall from LCA and DfE perspective. Int J Life cycle Ass 17(4):432–443
González-García S, Castanheira E, Dias AC et al (2013a) Using LCA methodology to assess UHT milk production in Portugal. Sci Tot Environ 442:225–234
González-García S, Castanheira E, Dias AC et al (2013b) Environmental life cycle assessment of a dairy product: the yoghurt. Int J Life Cycle Ass 18:795–811
González-García S, Bonnesoeur V, Pizzi A et al (2013c) Comparing environmental impacts of different forest management scenarios for maritime pine biomass production in France. J Clean Prod 64:356–367
Gottberg A, Morris J, Pollard S et al (2006) Producer responsibility, waste minimisation and the WEEE directive: case studies in eco-design from the European lighting sector. Sci Tot Environ 359(1–3):38–56
Guinée JB, Gorrée M, Heijungs R et al (2001) Life cycle assessment. An operational guide to the ISO standards. Centre of Environmental Science, Leiden

IDEMAT database (2001) Faculty of Industrial Design Engineering of Delft University of Technology. The Netherlands
Jungbluth N, Tietje O, Scholz RW (2000) Food purchases: impacts from the consumers' point of view investigated with a modular LCA. Int J Life Cycle Ass 5(3):134–142
Jungmeier G, Werner F, Jarnehammar A et al (2002) Allocation in LCA of wood-based products. Experiences of cost action E9. Part I. Methodology. Int J Life Cycle Ass 7(5): 290–294
Karlsson R, Luttropp C (2006) EcoDesign: what's happening? An overview of the subject area of EcoDesign and of the papers in this special issue. J Clean Prod 14(15–16):1291–1298
Koreneos C, Roumbas G, Gabari Z et al (2005) Life cycle assessment of beer production in Greece. J Clean Prod 13:433–439
Kurczewski P, Lewandowska A (2010) ISO 14062 in theory and practice—ecodesign procedure. Part 2: practical application. Int J Life Cycle Ass 15:777–784
Lewandowska A, Kurczewski P (2010) ISO 14062 in theory and practice—ecodesign procedure. Part 1: structure and theory. Int J Life Cycle Ass 15:769–776
Madival S, Singh SP, Narayan R (2009) Assessment of the environmental profile of PLA, PET and PS clamshell containers using LCA methodology. J Clean Prod 17:1183–1194
Marsh K, Bugusu B (2007) Food packaging—roles, materials, and environmental issues. Scientific status summary. J Food Sci 72(3):39–55
Mathieux F, Rebitzer G, Ferrendier S et al (2001) Ecodesign in the European electr(on)ics industry—an analysis of the current practices based on cases studies. J Sustain Prod Des 1(4):233–245
McDonough W, Braungart M, Anastas PT et al (2003) Applying the principles of green engineering to cradle-to-cradle design. Environ Sci Technol 37:434A–441A
Meyhoff Fry J, Edwards C (2011) Life cycle assessment of supermarket carrier bags: a review of the bags available in 2006. Report SC030148. Environment Agency, Feb 2011. 120 p
Moralejo-Gárate H, Palmeiro-Sánchez T, Kleerebezem R et al (2013) Influence of the cycle length on the production of PHA and polyglucose from glycerol by bacterial enrichments in sequencing batch reactors. Biotechnol Bioeng 110(12):3148–3155
Muñoz I, Rieradevall J, Domenech X et al (2006) Using LCA to assess eco-design in the automotive sector—case study of a polyolefinic door panel. Int J Life Cycle Ass 11(5):323–334
Nedermark R (1998) Ecodesign at Bang & Olufsen in product innovation and eco-efficiency. In: Klostermann J, Tukker A (eds) Twenty-three industry efforts to reach the factor 4. Kluwer Academic Publishers, Dordrecht
Peças P, Ribeiro I, Silva A et al (2013) Comprehensive approach for informed life cycle-based materials selection. Mater Des 43:220–232
PRé Consultants (2014) http://www.pre.nl
Ramani K, Ramanujan D, Bernstein WZ et al (2010) Integrated sustainable life cycle design: a review. J Mech Desing 132(9):1–15
Ribeiro I, Peças P, Silva A et al (2008) Life cycle engineering methodology applied to material selection, a fender case study. Mater Design 16:1887–1899
Ribeiro I, Peças P, Henriques E (2013) A life cycle framework to support materials selection for ecodesign: a case study on biodegradable polymers. Mater Des 51:300–308
Rivela B, Moreira MT, Bornhardt C et al (2004) Life cycle assessment as a tool for the environmental improvement of the tannery industry in developing countries. Environ Sci Technol 38(6):1901–1909
Rivela B, Moreira MT, Feijoo G (2007) Life cycle inventory of medium density fibreboard. Int J Life Cycle Ass 12:143–150
Ross S, Evans D (2003) The environmental effect of reusing and recycling a plastic-based packaging system. J Clean Prod 11:561–571
Ruhland K, Finkbeiner M, Griesser S et al (2004) Process and tools to support design for environment at Mercedes Car Group. In: 4th International Automobile Recycling Conference, Geneva (Switzerland)

Sanyé E, Oliver-Solá J, Gasol CM et al (2012) Life cycle assessment of energy flow and packaging use in food purchasing. J Clean Prod 25:51–59

Sanyé-Mengual E, Garcia Lozano ER, Oliver-Solá J et al (2014a) Eco-design and product carbon footprint use in the packaging sector. In: Muthu SS (ed) Assessment of carbon footprint in different industrial sectors, vol 1. Springer, Singapore

Sanyé-Mengual E, Pérez-López P, González-García S et al (2014b) Eco-designing the use phase of products in sustainable manufacturing. J Ind Ecol 18:545–557

Sanyé-Mengual E, Garcia Lozano R, Farreny R et al (2014c) Introduction to the Eco-design methodology and the role of product carbon footprint. In: Muthu SS (ed) Assessment of carbon footprint in different industrial sectors, vol 1. Springer, Singapore, pp 1–24

Siracusa V, Rocculi P, Romani S et al (2008) Biodegradable polymers for food packaging: a review. Trend Food Sci Tech 18:634–643

Smith J, Wyatt R (2006) Project inception: a performance brief approach. In: Proceedings of CRIOCM 2006 international research symposium on advancement of construction management and real estate, vol 1 and 2, pp 29–38

Spitzley DV, Keoleian GA, McDaniel JS (1997) Life cycle design of milk and juice packaging. Environmental Protection Agency. Center for Research Information. EPA/600/SR-97/082

Tischner U, Nickel R (2003) Eco-design in the printing industry life cycle thinking: implementation of eco-design concepts and tools into the routine procedures of companies. J Sustain Prod Design 3(1–2):19–27

Todd J, Brown E, Wells E (2003) Ecological design applied. Ecol Eng 20(5):421–440

Tukker A, Haag E, Eder P (2000) Eco-design: European state of the art. Part I: Comparative analysis and conclusions. An ESTO project report Prepared for the European Commission—Joint Research Centre Institute for Prospective Technological Studies, Seville. Available at: http://ftp.jrc.es/EURdoc/sps00140.pdf

Unger N, Schneider F, Salhofer S (2008) A review of ecodesign and environmental assessment tools and their appropriateness for electrical and electronic equipment. Prog Ind Ecol Int J 5(1–2):13–29

Van Hemel CG (1998) Ecodesign empirically explored D. U. of Technology, ed., TU Delft

Zust R, Wirnmer W (2004) Eco-design pilot—methods and tools to improve the environmental performance in product design. Tools Methods Competitive Eng 1&2:67–72

Organization Life-Cycle Assessment (OLCA): Methodological Issues and Case Studies in the Beverage-Packaging Sector

Alessandro Manzardo, Andrea Loss, Anna Mazzi and Antonio Scipioni

Abstract The management of packaging materials and their interactions with the environment is central to international debate. The reasons are manifold: packaging is essential to guarantee the good quality of the products they contain; its production can require the significant use of natural resources; and consumers' decisions are influenced by the environmental performances of packaging with particular reference to their management at the end of life. In this context, packaging companies has proved to be particularly interested in the application of environmental management and improvement tools such as life-cycle assessment. One of the latest developments of this methodology is its application at the organizational level, which was recently standardized in the ISO/TS 14072. Even if the interest around this topic is rapidly increasing and significant experiences are emerging (e.g., Organizational Environmental Footprint Programme of the European Union), no relevant applications have been published in the packaging sector. The objective of this chapter is to present the most relevant challenges in the application of the organizational life-cycle assessment for the packaging sector from the choice of the functional unit and the definition of the system boundaries to the choice on the aggregation approaches and the assessment of environmental impacts. Such issues will also be presented from a practical perspective presenting relevant case studies and lessoned learned in the beverage-packaging sector.

Keywords Organizational life cycle assessment · Beverage-packaging · Organizational system boundaries

---

A. Manzardo · A. Loss · A. Mazzi · A. Scipioni (✉)
Department of Industrial Engineering, CESQA C/O University of Padua,
Via Marzolo 9, 35131 Padua, Italy
e-mail: scipioni@unipd.it

S.S. Muthu (ed.), *Environmental Footprints of Packaging*,
Environmental Footprints and Eco-design of Products and Processes,
DOI 10.1007/978-981-287-913-4_3

1 The Evolution of Organizational Life-Cycle Approaches in International Standards

After the increase of awareness of the international community of environmental issues, a growing number of organizations are adopting different environmental management tools in order to assess, monitor, and reduce the environmental impacts generated from their activities. The packaging sector proved to be particular sensitive to this topic (Scipioni et al. 2010); many companies in fact have assessed the environmental performances of a significant share of products according different environmental schemes, in particular life-cycle assessment (LCA), according to ISO 14040 and ISO 14044 (ISO 2006a, b) and environmental product declaration (EPD) according to ISO 14025 (ISO 2006c). LCA methodology has been a quantitative tool to support the decision-making process toward environmental sustainability. Although LCA was originally developed for products, recent scientific developments have demonstrated that the benefits of the life-cycle approach can be extended to the environmental assessment of more complex organizations and their value chain (UNEP/SETAC 2015).

The first applications of the life-cycle approach at the organizational level were focused on the quantification of climate change impacts of companies and their value chain (UNEP/SETAC 2015). This issue was in fact the first to be treated at an organizational level by standards such as the Greenhouse Gas Protocol (WRI and WBCSD 2001) and the ISO 14064-1 (ISO 2006d).

The GHG Protocol presents probably the first attempt to apply the life-cycle approach at organizational level. Published in its first version in 2001 it is the result of a joint initiative of the World Resources Institute (WRI) and the World Business Council on Sustainable Development (WBCSD). It specifies requirements and guidelines for the quantification of greenhouse gas emissions that directly arise from processes under the control of organizations and their value chain. ISO 14064-1 (ISO 2006d) was published in 2006 incorporating many of the concepts presented in the GHG protocol including the possibility to adopt a life-cycle perspective in monitoring greenhouse gas emissions. The increasing importance of adopting life-cycle approaches at the organizational level has also been confirmed by the recent publication of the ISO/TR 14069 (ISO 2013), which gives clear examples on how to apply ISO 14064-1 outside of the physical boundaries of an organization toward a more comprehensive life-cycle perspective. ISO 14064-1 is currently under revision; the actual mandate of the ISO indicates that the future version of this standard will probably strengthen the life-cycle approach. Other organizational approaches that allow for the quantification and management of environmental impacts are related to the establishment of environmental management systems (EMS) according to ISO 14001 and the European Eco-Management and Audit Scheme (EMAS) (EC 2009). Even if not directly mentioned, the life-cycle approach can be recognized in the way suppliers are considered in the management of environmental impacts related to the activities of organizations. The new version of ISO 14001, in particular (ISO 2015a), has further strengthen the management of impacts using a life-cycle perspective. The increasing interest for the application of life-cycle approaches at the organizational level has finally been confirmed by the publication of recent ISO standards such as ISO 14046 on Water Footprint (ISO 2014) and ISO/TS 14072 on Organizational LCA (OLCA) (ISO 2015b); the launch of institutional initiatives, such as the recommendation of the 9th of April 2013 on Organization Environmental Footprint of the European Union (OEF) (EU 2013); and the Flagship project on LCA of organizations (O-LCA) of the UNEP-SETAC Life Cycle Initiative (UNEP/SETAC 2015). The new ISO standard on Water Footprint, published in 2014, includes specific requirements that focus on the quantification of potential environmental impacts related to water of the activities of organizations from a life-cycle perspective. It is the first ISO standard on LCA that directly mentions organizational approaches and that presents the integration of products and organizational perspectives. Considering the importance of water use in paper-based packaging, this standard will surely find wide applications in the packaging sector (Manzardo et al. 2014). Its publication anticipated ISO/TS 14072 (ISO 2014d), which was released in January 2015. This standard provides recommendations and requirements specifically designed to facilitate a more effective application of ISO 14040 and ISO 14044 (ISO 2006a, b) to organizations. The document describes how to adapt the requirements of product LCA to organizations and the potential benefits that doing so can bring. The organization environmental footprint (OEF) method has been developed by the Joint Research Centre of the European Commission, which has recognized the importance of assessing environmental potential impacts related to the life cycle of the activities of an organization. The OEF is a multi-criteria measure of the environmental performance of a product-providing organization from a life-cycle perspective (Pelletier et al. 2014). Although the OEF can be seen as a type of organizational LCA, it is not completely in line with some principles and requirements of ISO 14040 and ISO 14044 (Finkbeiner 2014; Galatola and Pant 2014) by specifying a cut-off criteria, recycling formula for end-of-life, and the default set of impact categories and indicators (Finkbeiner 2014). The efforts of the UNEP/SETAC Life Cycle Initiative recently resulted in the publication of "Guidance on Organizational Life Cycle Assessment" (UNEP/SETAC 2015), which introduces the methodological framework of O-LCA and presents several example to facilitate its application and is in line with the contents of ISO/TS 14072. This initiative is now entering a testing phase where a number of companies will apply the published guidelines and will provide feedback for future improvements.

Considering the different range of guidelines and standards published on the topics, in this chapter it was decided to present the application of OLCA to the packaging sector adopting the language and the perspective of ISO/TS 14072. Where relevant, differences among the different approaches will be presented. The general characteristics of the above mentioned references are reported in Table 1.

Table 1 General characteristics of standards and guidelines that adopt a life-cycle approach

Standard/guideline/Technical Specification	Life-cycle approach	Goals	Foreseen application
GHG Protocol: corporate accounting and reporting standards (WRI and WBCSD 2001)	Application of the life cycle approach can be performed with reference to the so-called "scope 3 emissions." These emissions arise from processes related to the activities of organizations but not under their direct control (e.g., raw materials and ancillary materials production, end-of-life processes)	Provide guidance to companies in creating a true and fair account of their GHG emissions	GHG protocol intended to support accountancy and disclosure for both internal use and a range of external applications
		Support participation in voluntary and mandatory GHG programs	Managing GHG risks and identifying reduction opportunities
		Increase consistency and transparency in corporate GHG reporting	Public reporting and participation in voluntary GHG programs
			Participating in mandatory reporting programs
			Participating in GHG markets
			Recognition for early voluntary action
			For use by businesses and other private or public organizations
ISO 14064-1 2006 Greenhouse gases. Part 1: specification with guidance at the organizational level for quantification and reporting of greenhouse gas emissions and removals (ISO 2006d)	Application of the life-cycle approach is related to the so-called "other indirect emissions". These emissions arises from processes related to the activities of an organizations but not under their direct control (e.g., raw materials and ancillary materials production, end-of-life processes)	Provide principles and requirements for organizational design, development, management and reporting of GHG emissions	Organizational design, development, management and reporting of GHG emissions for the purpose of corporate risk management, voluntary initiatives, GHG markets, or regulatory reporting

(continued)

Table 1 (continued)

Standard/guideline/Technical Specification	Life-cycle approach	Goals	Foreseen application
ISO/TR 14069 2013 Greenhouse gases. Quantification and reporting of greenhouse gas emissions for organizations. Guidance for the application of ISO 14064-1 (ISO 2013)	The life-cycle approach is presented in several case studies with reference to the category "other indirect emissions" of ISO 14064-1	Detailed guidance for the implementation of ISO 14064-1	Allow for the analysis of greenhouse gases of organizations outside the physical boundaries eventually adopting a life-cycle approach
ISO 14046 2014. Environmental Management. Water Footprint. Principles, requirements and guidelines (ISO 2014)	Life-cycle perspective is one of the principles of this standard	Provide principles requirements and guidelines related to water footprint assessment of products, processes, and organizations based on life-cycle assessment (LCA)	Quantification of potential environmental impacts related to water of products, processes, and organizations
			Identify opportunity to reduce water use and impacts within organizations, related activities, and life cycles
			Facilitate water efficiency and optimization of water management
			Report water-footprint results to stakeholders and decision makers
Recommendation of the Commission of the European Union on the use of common methods to measure and communicate the life-cycle environmental performance of products and organizations. 9th of April 2013 (EU 2013)	Life-cycle perspective is at the ground of the recommendation	Provide a series of comprehensive and detailed technical guidelines for the conduction of a OEF stud within the European voluntary framework	Quantification and reporting of environmental information to be used within the European Union also with reference to voluntary schemes launched by the member states

(continued)

Table 1 (continued)

Standard/guideline/Technical Specification	Life-cycle approach	Goals	Foreseen application
ISO/TS 14072 Environmental management. Life-cycle assessment. Requirements and guidelines for Organizational Life Cycle Assessment (ISO 2015b)	Life-cycle perspective is one of the principles of this standard	Provide additional requirements and guidelines for an effective application of ISO 14040 and ISO 14044 to organizations	Quantification of environmental performances and impacts of an organization
			Performance tracking over time
			Reporting of environmental performances to the public
			Support to decision makers
UNEP/SETAC Guidance on Life Cycle Assessment (UNEP/SETAC 2015)	Life-cycle perspective is a core principle of this guidance	Presents benefits and methodological challenges of the application of LCA to organizations	Quantification of environmental performances and impacts of an organization and its supply chain
			Performance tracking over time
			Reporting of environmental performances to the public
			Support to decision makers
			Reduce operational costs
			Understand risks and impact reduction opportunities

2 Methodological Aspects of the Application of OLCA

According to the definition of ISO/TS 14072 (ISO 2014d), OLCA consists of the compilation and evaluation of inputs, outputs, and potential environmental impacts of activities associated with an organization either as a whole or a portion thereof adopting a life-cycle perspective. Aligning with other ISO LCA product-based standards, the general structure of an OLCA study (Fig. 1) is based on four phases: (1) goal and scope definition, (2) inventory analysis (LCI), (3) impact assessment (LCIA). and interpretation. However, considering the specificity of organizations, new requirements can be identified in the content of the technical specification.

According to ISO 14072 (ISO 2014d), an organization is defined as a person or group of people that has its own functions with responsibilities, authorities, and relationships to achieve its objectives.

The definition of the goal and scope of the study within ISO/TS 14072 requires mainly the identification of the intended application, the reasons for carrying out the study, and the intended audience (e.g., internal or external communication and types of stakeholders). Often the goals are linked with the intention of identifying impact-reduction opportunities along the value chain, tracking performance over

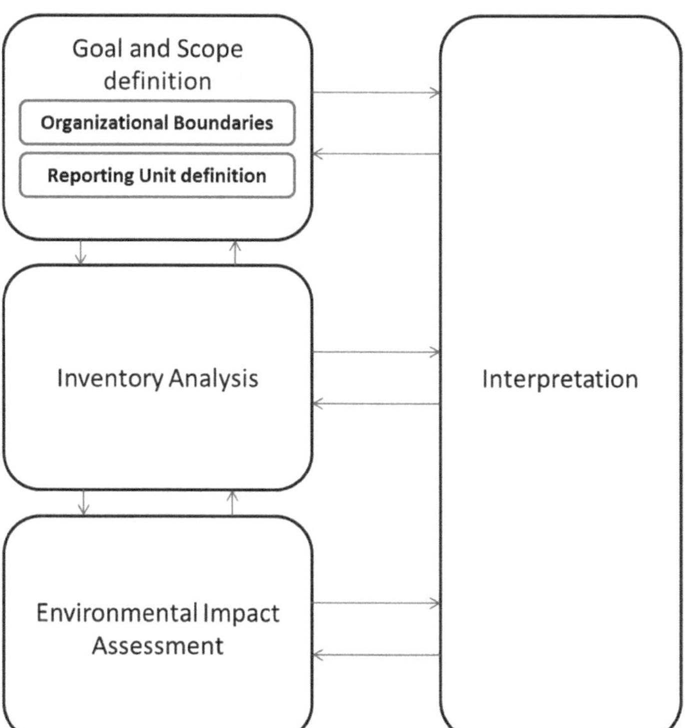

Fig. 1 Life-cycle assessment framework adapted from ISO 14040 (ISO 2006a)

time, or improving knowledge, control, management, and transparency of operations (Table 1). These requirements are perfectly in line with ISO 14040 except for the issue of comparative assertion. This opportunity is excluded from ISO/TS 14072 because during development of the standard, the difficulty in comparing different organizations, even if they belong to the same sector, became clear. When the goal of the study is fully clarified, the system boundaries of the study can be determined. This phase is probably the one where the major difficulties were encountered in the process of adapting the principles and requirements of product LCA standards to OLCA standards (Martinez-Blanco et al. 2015); in fact, whereas a product and its functions can be easily identified, this is not the case when looking at organizations. This is particularly relevant in the case of complex organizational structures that may deliver several products and services and may share other companies' revenues (DEFRA 2013). For these reasons, the definition of scope in OLCA should start with the clear identification of the organization under study, e.g., its products, operations, facilities, and sites. Borrowing a concept expressed in ISO 14064-1 (ISO 2006d), the OLCA starts with a definition of the boundaries of the organization and the definition of the consolidation method to be adopted for aggregating the results of the study. Identification of organizational boundaries is part of the process of system-boundaries definition. However, the two concepts are not the same: If the former is related to the life-cycle stages and activities to be included, the latter clearly answers the need to identify who is the organization under study and therefore which sites, installations, and companies should be considered in the study. According to ISO 14072, two approaches can be adopted to identify the organizational boundaries and therefore consolidate the inventory data and impacts at a facility and sites level: the first one is the financial and operational control, and the second one is the equity share. If the first approach is chosen, all of the activities under direct control of the organization from an operative and/or financial perspective are considered. This approach is usually adopted by small- and medium-sized organizations with simple financial structure and limited participation in other companies. The main benefit of this method is that only the units directly influenced by the organization are included in the study, and therefore the collection of data and the implementation of potential improvements detected during OLCA are easier. However, these approaches do not fully reflect the financial risks and rewards compromising financial-risk management (Martinez-Blanco et al. 2015). When adopting the equity share approach, the organization assesses the impacts of processes and physical units from respective facilities and activities according to its share of equity interest. This approach is generally adopted in the case of companies with complex structures. In fact, it is more straightforward when the organizational structure is complex and facilitates financial management by reflecting the full financial risks and rewards and is thus OLCA is less subject to interpretation (Martinez-Blanco et al. 2015). However it must be noted that this choice could be particularly expensive and time consuming; in fact, in this case activities with a very small share over which the company has no control are also be considered in the study. In the case, the organization fully owns and controls all of its activities, and the organizational boundaries will be the same regardless of what consolidation method is chosen.

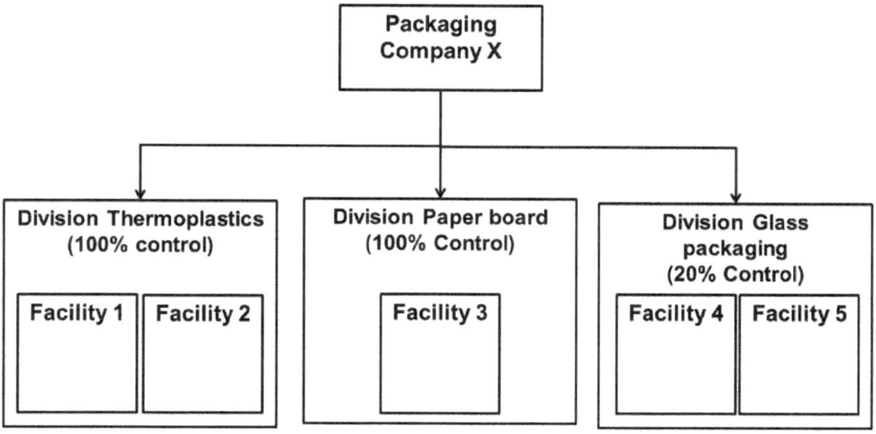

Fig. 2 Examples of potential organizational boundaries

Figure 2 represents an example of organizational boundaries adopting the two consolidation methods. For example, packaging company X fully owns the activities of a thermoplastics division, a paper-board division, and their related facilities. It also has a 20 % share of a glass-packaging division, but it has no financial or operational control over the facility's activities. If packaging company X adopts the operational and functional approach method, it should consider only all of the activities of the divisions it fully owns. If packaging company X chooses the equity share approach, this should also include 20 % of the impacts generated from the glass-packaging division and its related facilities.

This step is the ground for the definition of the so-called "reporting unit," which—in the case of OLCA—substitutes for the concept of functional unit. The reporting unit is the performance expression of the organization under study to be used as a reference (ISO 2014d). Performing a parallel comparison with the product LCA standards, the amount of impact in the OLCA study is the "functional unit." Examples of reporting unit are (1) in the case of a company that produces only beverage cartons the reporting unit would be the total amount of polylaminated carton produced in the reference year or (2) in the case of as organization under study with a large variety of products, the reporting unit could be identified as a certain amount of revenue coming from the production and sale of its products.

ISO/TS 14072 allows flexibility in the definition of the reporting unit allowing for different levels of assessment. From this perspective, through the definition of the scope of the study, the organization may decide to focus on either the organization as a whole or its portions such as business divisions, brands, regions, or facilities (UNEP/SETAC 2015) (Fig. 2). In this latter case, the OLCA study refers to a subset of the organization. The application of the international standard to segments or selected parts of an organization can be performed if it is properly justified (Martinez-Blanco et al. 2015) (Fig. 2). This perspective can be supportive to packaging companies with a complex product portfolio and several business units; in fact these organizations can plan to start a pilot assessment on a part of the organization and move to a more complete OLCA in the future. This is also useful for companies that produce their products at essentially independent sites for diverse sectors (e.g., packaging products and food products). Furthermore, it is important to take into account that in some case, assessment of the whole organization may be financially unacceptable due to limited knowledge and/or control over operations (Martinez-Blanco et al. 2015). Organizations that perform a subset analysis, however, should be aware that this probably will not consider part of its value chain and therefore limits it's the definition of its impact-reduction strategy.

Flexibility is allowed also in the definition of life-cycle stages and process units to be included in the study when the exclusions are adequately justified. For instance, this could be the case of the use stage of beverage packaging such as beverage cartons and polyethylene (PET) bottles containing fluid that needs to be stored in a refrigerated environment (e.g., different types of milk); in fact, in this case the energy used for refrigeration can be relevant for the content but not relevant for the containers.

In the definition of scope, other important new issues should be considered at the organizational level. This is the case of the so called "reference period." which consists of the period decided by the organization serve as the first assessment of the organization's environmental impacts. This is a key choice when the organization is willing to undergo performance tracking over time, but it also contextualizes results for organizations where processes and productions can change many times during limited period of times. Considering the requirements of the standard, a company can decide the length of the reference period. This could be a year (Scipioni et al. 2010) or a season when seasonality is a relevant issue for the company under study. Therefore, the results of an OLCA study are time-dependent.

The other requirements of the OLCA scope definition are perfectly in line with ISO 14040 (ISO 2006a) including the system boundary definition, the allocation procedures, the impact-assessment methodologies and types of impacts, the data-quality requirements, the interpretation to be used, the assumptions, the limitations, and the type of critical review and reporting.

Figure 3 describes the case of a packaging company that owns the 100 % of two subdivisions (Thermoplastics and Paper Board) and their related sites. When implementing OLCA, the company can decide to consider a subset of the whole organization. For example, when the goal of the study is to assess the environmental impacts of a thermoplastics division according to the organizational boundaries of packaging company X, the boundaries of the analysis should include the activities of facility 1 and facility 2 as well as the related life-cycle processes. In the case of products realized by facility 2 only, the distribution and use stages were considered, but they turned out to be out of the control of the reporting organization.

It is important to point out that the system boundary is directly linked with the organizational boundary definition, the reporting unit, and the reference period chosen for the assessment. Pelletier et al. (2014), highlights that often the resources use and the emissions linked to upstream processes (e.g., raw material

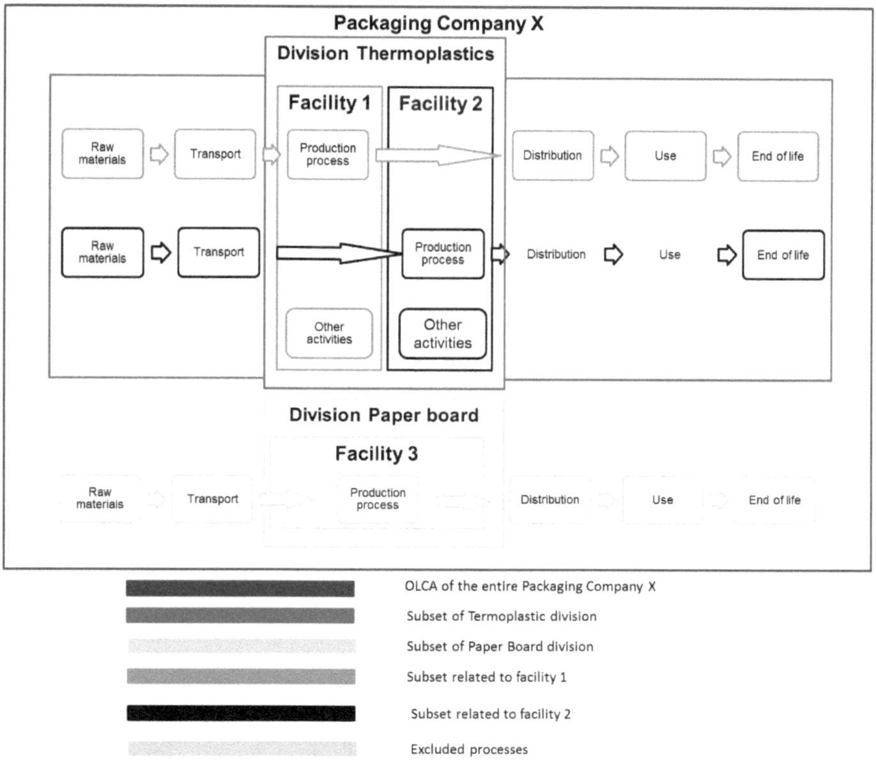

Fig. 3 Examples of potential subsets to be assessed

extraction, processes of material production, supply transport processes) and/or downstream processes (e.g., distribution and end-of-life processes) can be as or even more determinant of the overall environmental profile of the reporting organization.

This fact was confirmed in several case studies: In the study of Scipioni et al. (2010, 2012), where a beverage carton company is considered, impacts coming from the supply chain and end-of-life processes contributed to approximately 95 % of final impacts on climate changes.

According to ISO/TS 14072, the exclusion of life-cycle stages is allowed only in the case that such processes cannot be influenced by the organization and/or present irrelevant environmental impacts. Another important step of goal and scope definition is the selection of environmental-impact categories to be included in the study. It is necessary to decide whether the environmental impacts of the organization are assessed at the mid-point or end-point level. In either case, the selection of impact categories, category indicators, and characterization should be justified and referenced or described (ISO 2014d). A packaging company deciding to perform OLCA could decide to use published sectorial product category rules as references for the first selection of impacts categories to be considered; the initiative on OEF of the European Union will probably provide this information in the future for packaging companies that adhere to the pilot programme for the definition of such PCRs. This selection, according to the iterative procedure of the OLCA, could be refined after the interpretation of the results phase, which is the last step of an OLCA.

Considering the packaging sector and, in particular, paper-based products where water use can be relevant in processes such as the growth of trees and paper production in paper mills, it is recommended to consider the latest developments in impact-assessment methodologies related to water (Kounina et al. 2013) and consider both consumptive and degradative water use (Bayart et al. 2013).

The second phase of an OLCA study is the life-cycle inventory phase. Inventory analysis involves data collection and calculations to quantify relevant inputs and outputs of all the activities that were considered in the organizational boundaries and in the system-boundaries definition. The inventory covers data regarding all inputs (e.g., energy, materials, water) and outputs (e.g., products, waste, emissions to air, soil, and water). The type of data used, the quality of the data, and the data sources should be transparently reported in the study. In general, all of the data can be divided in three different subsets: data regarding upstream processes, data regarding processes directly performed by the "reporting organization," and data related to downstream processes. The first subset of data in the case of a packaging company can be linked, but not limited to, processes such as the extraction and/or production of crude oil, wood, and water, fuels, ancillary materials (chemicals) and goods (e.g., intermediate products), outsourced services (e.g., marketing, legal, information technology [IT] and logistic services), capital equipment (e.g., injection molding machine), extraction, the production and distribution of purchased electricity, and steam and energy imported by the organization. The second subset is related to processes that are directly performed by the organization under study. For example, in the case of a packaging company, these could be the generation of energy resulting from the combustion of fuels in stationary sources (e.g., boilers, furnaces, and turbines); the mulching and leaching during bleached paper production; the transportation of materials, intermediate products, products, and waste; and the disposal and treatment of solid and liquid waste. Finally, the third subset is linked to downstream processes such as transportation and distribution of finished products, the processing and storage of products provided to the client, use or consumption of the provided goods, and end-of-life (EoL) treatment of packaging after use. Different data sources can be considered during inventory analysis; in order of preference, these are site-specific primary data, secondary data, and tertiary data (ISO 2006a). Examples of sources of site-specific primary data are for example-consumption data, bills, emissions directly measured and reported to competent authorities (e.g., air emissions, water emissions), mass balance or stoichiometry, and composition of waste and products. Common sources of secondary data are data bases such as sectorial or governmental datasets, scientific articles, and life-cycle inventory databases such as Ecoinvent

(Weidema et al. 2013). Finally, tertiary data are linked to estimation procedures or expert opinions.

According to ISO/TS 14072, the collection of inventory data can be performed according to three different procedures that result in different levels of aggregation of data. The first one is a top-down approach: In this case, data are collected looking at the whole organization without referring to a specific production or service delivered. This choice is usually preferred because data such as energy, raw materials purchase, waste production, etc. are generally accounted for in organizations without referring to a specific product or process. This approach, however, limits the possibility of the company to identify specific hot spots and/or areas of improvement related to specific processes and/or products. The second approach is a bottom-up approach: This is the case when data on specific products are available and aggregated together to cover the whole spectrum of the activities considered in the organizational and system boundaries. This procedure can be more complex because the details of data are greater; however, it allows for a better understanding of the contribution that each product and process brings to the overall impacts related to the organization under study. Finally, the hybrid approach is a combination of the previous two approaches (Martinez-Blanco et al. 2015).

Concerning the life-cycle inventory phase, no substantial changes with respect to ISO 14040 families have been introduced in the adaptation process. Mainly, some recommendations are provided by ISO/TS 14072 regarding the collection of data, the application of cut-off criteria, and the allocation procedure. In fact, at an inventory level, is not recommended to aggregate the OLCAs of the supply chain when the company does not own the whole supply chain and/or buy all the production of its suppliers.

In life-cycle inventory phase, additional requirements must be considered in the case when water is a relevant aspect to be considered in the OLCA study. In fact, ISO 14046 does not allow the aggregation of data at the inventory level especially in the case when the sites of the organization are located in different regions with different water availability (ISO 2014). This is valid also in the case when the products coming from a supplier are produced in different locations. This situation is common in many sectors and can also be relevant for packaging companies.

The third phase of OLCA study is the stage of impact assessment. It consists of the same steps of product LCA: classification and characterization, which are mandatory; normalization; and aggregation and weighting, which are optional (ISO 2006b). In this phase, the impact-assessment categories determined in the definition of goal and scope are considered, and impact-assessment methods are applied.

As previously mentioned, the interpretation of results is the last phase of OLCA In this phase, the findings of the inventory analysis and the impact assessment are considered together. The interpretation phase should deliver results that are consistent with the defined goal and scope and should reach conclusions, explain limitations, and provide recommendations. It is analogous to that of product LCA, meaning that recommendations and requirements for the latter are applicable to the former.

3 Possible Applications and Expected Benefits of OLCA in the Packaging Sector

Applications of OLCA are manifold and may differ between different types and sizes of organizations. In the case of the packaging sector, considering the interest for environmental issues, the following general application can be identified. A first application could be (1) the use of OLCA studies along with the implementation of EMS according to ISO 14001; (2) during the initial environmental review, OLCA can support the quantification of environmental impacts and identify environmental significant aspects with a focus on the organization and its supply chain; (3) during the planning phase, OLCA can support the quantification of targets along with the definition of specific objectives such as in the design of a green supply chain and identification of green suppliers; (4) during the check phase, OLCA could be useful to quantify the environmental performance of the organization and its supply chain and determine if the targets were successfully achieved within the company and among suppliers; and (5) OLCA results can be used during the management review to support the process of continuous improvement. Considering that ISO 14001 is currently under revision and that the life-cycle approach will be even better-detailed in the new version, it can be argued that OLCA use within organizations implementing EMS will increase in the future (ISO 2015a). Implementing OLCA within EMS is expected to bring benefits to the organization, e.g., OLCA can increase the knowledge on internal processes and improve the understanding operations along the value chain (UNEP/SETAC 2015).

Another application of OLCA is to support the strategic decision-making process. OLCA is in fact a comprehensive assessment of environmental impacts; therefore, it can help in prioritizing actions to reduce the environmental impacts of products as well as the operations of the organization, thus avoiding the practice of environmental burden-shifting. When the OLCA is applied as a strategic tool, it supports the identification of environmental hot spots at different levels, e.g., between inputs and outputs, processes, business divisions, brands, regions, or facilities. In doing so, the organization identifies which areas are at risk and where opportunities exist for resource efficiency and emissions mitigation regardless of whether they occur within the organization's boundaries or upstream or downstream in the value chain. Likewise, OLCA helps the adoption of more environmentally friendly management and eco-innovation approaches in the organization and along the value chain.

According the recent UNEP-SETAC guideline of OLCA (UNEP/SETAC 2015), one of the most relevant application is related to performance tracking overtime. This application is also explicitly mentioned in ISO/TS 14072. When an organization wants to perform performance tracking, additional care should be taken. First of all, it is important to determine a base period to be used as reference for comparison. This is to be considered a strategic decision. A company can decide to use a specific period coming from applicable regulation (e.g., Kyoto targets) (UNFCC 1997) or programme (CDP 2014a, b), or it can choose a period depending on data validity and availability. In a performance-tracking application, it is important to document any changes occurring within the organization. If these changes are relevant, a new assessment of the base year to reflect such changes is needed (Scipioni et al. 2010). The application of OLCA as a tracking tool is encouraged because it is an appropriate framework for tracking environmental performance over time, both at the inventory and impact levels, in a similar fashion to how organizations use financial and activity data. Performance tracking responds to multiple organizations' necessities. For example, it helps in tracking improvements in the environmental performance of the organization in reference to a certain internal or external target.

Considering the market value of packaging, probably one of the most relevant applications of OLCA is the reporting and communication of the environmental impacts of an organization over a given period of time. Even if communication is not included in ISO 14072, OLCA is a scientific-based approach that is also based on the principles of transparency and comprehensive communication. The results of an OLCA study can therefore be used to support communication and information to stakeholders, consumers, investors, authorities, and the general public. One of the most relevant targets of communication based on OLCA results could be to increase the company's reputation in the market and give evidence of its sustainability. For instance, if performance tracking is performed, a company could claim its performance and achievements over time (Scipioni et al. 2010).

In this "contest," to achieve optimal performance, it is important to remember that OLCA according to ISO 14072 is not intended for comparative assertions to be disclosed to the public. A company can decide to apply OLCA for internal benchmarking (e.g., compare owned sites that perform the same activities); however, comparison with competitors can be meaningful or weak at any given time (UNEP-SETAC 2015). This is the result of the difficulties in setting the conditions for comparison (e.g., the same reporting unit). Initiatives such as the Environmental Footprint of the European Union have been studied to overcome these limitations (EU 2013). The OEF process is actually in the stage of pilot applications to determine sectorial rules to support the comparison and benchmarking of environmental performances of companies belonging to the same sector. At the moment, however, only two sectors are under investigation—copper production and retail—and no packaging OEF rules are expected at the moment (EU 2013).

4 Case Studies of the Applications of OLCA Approaches in the Packaging Sector

OLCA is considered to be one of the most important emerging application of the life-cycle approach (Hellweg et al. 2014), which only recently has been standardized (ISO 2015b). For this reason, case studies of its application refer to pilot studies and are usually limited to one impact category such as climate change

(Scipioni et al. 2012); moreover, these experiences are focused on a limited number of sectors (UNEP-SETAC 2015). One field that has proved to be particularly interested in this method is packaging with specific reference to foods and beverages. The reasons for this interest are manifold. The food industry is one of the world's largest industrial sectors and a main energy consumer (Manfredi et al. 2015). Food production, preservation, and distribution indeed consume a considerable amount of energy, which causes resource depletion and pollutant emissions (Roy et al. 2009). Packaging is a fundamental element for almost every food and beverage product and thus is a vital source of environmental burden and waste. Packaging isolates food from factors affecting loss of quality such as oxygen, moisture, and microorganisms, and it provides cushioning protection during transportation and storage (Roy et al. 2009). The packaging of food products presents considerable challenges to the food and beverage industry, and minimizing packaging and modifying both primary and secondary food packaging present an optimizing opportunity for these industries (Henningsson et al. 2004; Hyde et al. 2001). In fact, the production stage of a packaging system has been reported be the principal cause of major environmental impacts. Furthermore, packaging utilization in the food and beverage sector, together with trends of increased consumption of packaged products, contributes to a growing volume of packaging waste (WPO 2008; EUROSTAT 2010).

To assess the environmental sustainability and reduce environmental impacts caused by the food and beverage sector, mainly from packaging production, LCA methodology (ISO 2006a, b) has been widely applied in several case studies (Manfredi and Vignali 2015; Notarnicola et al. 2012; Roy et al. 2009). Moreover, to assess and improve the environmental sustainability of all of their activities without focusing only on one specific product, a few organizations, have used environmental-management tools at the organizational level that can be considered the first applications of OLCA procedures limited to one specific impact category.

For instance, in Scipioni et al. (2010), ISO 14064-1 (2006d), for the quantification and reporting of greenhouse gas emissions and removals, a world leader beverage-carton company adopted a life-cycle approach. The objective of the study was to support the company in implementing a performance-tracking system of the impacts generated during the life cycle of its activities. The system was designed to allow the company to quantify its performances related to climate change and monitor the achievement of a greenhouse gas emission–reduction target beyond its physical boundaries. In this case study, the requirements of ISO 14040 (2006a) were used in the definition of the goal and scope and in the identification of downstream and upstream processes to be included in the performance tracking system. The approach of the operational control was used in the definition of the organizational boundaries; therefore, this approach considered all of the activities under the direct control of the organization and their respective life-cycle processes. The reporting unit (formerly the functional unit as reported in the study) was determined as the total quantity of finished products produced in 2004. The company in fact produces one type of beverage carton material that can be sold in different formats.

Inventory analysis was performed according to ISO 14044 (2006b) and therefore also complied with the current version of ISO 14072 (2015b). In this case, a bottom-up approach was followed collecting data on all inputs and outputs aggregated at the organizational level.

The results show that GHG emissions, such as raw materials production (>60 %), are mainly generated by processes not under the direct control of the organization. Another large contribution is related to the products' end of life, accounting for approximately 15 % of the overall emissions, and transports, accounting for approximately 14 % of the overall emissions. Impacts from production activities were limited and depended on energy consumption, which accounts for 8 % of the overall emissions. These results were used by the company to identify solutions to reduce its impacts on climate change of activities under its direct control. They also helped the company to set the basis for the improvement of environmental performances of downstream and upstream processes related to climate change.

Following the outcomes of this research, Scipioni et al. (2012) set up a new methodological framework for the integration of LCA (ISO 2006a, b) and ISO 14064-1 to align the monitoring and management of the GHG emissions determined at the organizational and product levels.

The study presents a procedure to determine how decision-making at an organizational level affects the carbon footprint of its products. This study was a first attempt to conduct performance tracking at the product level. It proved the existence of a relationship between organizational and product approaches for impact assessment. This relationship has been confirmed also by UNEP_SETAC initiative on OLCA (UNEP/SETAC 2015).

5 OLCA of PET- and Glass-Bottled Mineral Water: A Case Study

5.1 Introduction

In this section of the chapter, a first application of OLCA according to ISO 14072 with specific focus on packaging is presented. The organization under study is San Benedetto S.p.A (hereafter San Benedetto), an Italian company leader in the food–beverage sector. San Benedetto was the first company in Italy to bottle water in PET containers. Currently the company owns 11 sites around the world with a sales network covering approximately 100 countries on 5 continents.

5.2 Goal and Scope Definition

San Benedetto has always considered environmental themes in its business and operations, and sustainability has become important for its long-lasting competitive advantage. In line with its environmental policy, the company started to adopt life-cycle approaches to minimize the impacts of its products and processes. In 2014, the company launched a new project to extend the control of its environmental performances by adopting an OLCA approach. The goal of the study was to quantify the environmental impacts related to the division concerned with bottling of mineral waters from San Benedetto spring located in Scorzè (Venice). The intended applications of the OLCA for San Benedetto were to provide a system for tracking its environmental-performance activities and to develop a pilot model to be extended in the future to the other divisions and sites of the company. The company fully owns the operations and activities located in the division under study; the consolidation method of the financial and operational control was chosen to determine the organizational boundaries. According to ISO/TS 14072, all of the activities and related life-cycle processes of the segment (division) of the organization under study were therefore considered (Fig. 4). This included the extraction, transformation, and transport of raw and ancillary materials from different suppliers, the processes that directly take place in the company, the

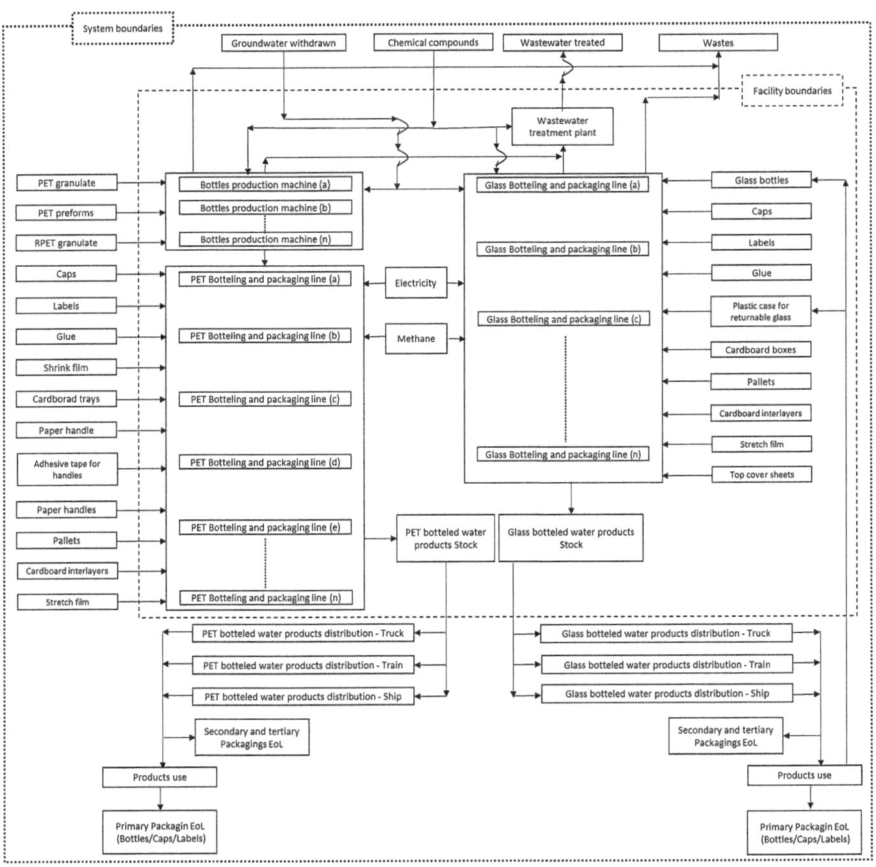

Fig. 4 Life-cycle processes of the division under study

Table 2 Product portfolio of the division under study

Product line	Format (l)	No. of product references
Glass	1.0	33
	0.75	20
	0.50	19
	0.25	11
PET	2.0	8
	1.5	50
	1.0	17
	0.75	6
	0.50	50
	0.33	9
	0.25	8

distribution of finished products, the use stage, and the end-of-life operations. At the same sites, other products are produced by two other different divisions: the soft drink division and the Guizza spring water division.

In the division under study, PET bottles are produced, and water is bottled in different formats: PET bottles and glass bottles (Table 2). The company directly operates the process of PET-container production and glass-container recycling. Figure 1 reports the representative life-cycle processes of a generic PET-bottled water and a generic glass-bottled water.

According to the goal of the study, the reporting unit considered was the overall volume of water (1.028.995 m^3) drawn by San Benedetto spring and bottled in PET and glass containers in the year 2013. This year was chosen as the reference year for the performance tracking of the company's environmental performances. According to the guideline from UNEP-SETAC on OLCA, Table 2 reports the product portfolio produced in the reference period. This practice improves the capability of interpretation of results during performance tracking.

References belonging to the same product category (i.e., PET- and glass-bottled water) differ from physical characteristics such as the container's weight, the percentage of recycled PET, caps and labels, secondary packaging (trays or cartons), and, finally, differences in tertiary packaging such as pallet and extensible film.

The following impact categories were considered to be relevant for the impact assessment: climate change, ozone depletion, terrestrial acidification, terrestrial ecotoxicity, water depletion. Aquatic eutrophication, aquatic ecotoxicity, aquatic acidification, human toxicity, photochemical oxidation, particulate-matter formation, ionizing radiation, metal depletion, fossil depletion, and water availability. The environmental performances were assessed adopting methods from ReCiPe 2008, Europe Midpoint (H) (Goedkoop et al. 2013), "IMPACT 2002+" (Joliet et al. 2005), and Boulay et al. (2011). The choice of integrating the two methods is related to the desire to have a more comprehensive analysis of impacts related to water.

5.3 Life-Cycle Inventory

In this case study, to verify the consistency of the data collected and the mass and energy balance, the life-cycle inventory phase was performed firstly adopting a bottom-up approach and then applying a top-down approach.

In the bottom-up approach, data were product oriented and followed the consolidate practice of LCA traditional-product studies. Data were collected separately for each product reference of the product portfolio. Material inputs are related to different aspects, e.g., primary, secondary, and tertiary packaging materials; chemical compounds using for the sanitization and cleaning of the installations; and chemical compounds used for wastewater treatment. Energy inputs are related to electricity and methane-gas consumptions. Other information collected at the product level is related to distances traveled for the distribution of every product in terms the function of different means of transport (truck, train, and ship). All of these data were mapped within the facility to identify the processes and machineries involved in the production of each single product reference. PET containers are produced in a department that operates 38 different machineries, which presents technological and performance differences. The bottles produced are filled and packed in 19 different bottling lines. Tools for data management were specifically created to simplify and efficiently support the extensive data collection. The main challenges during site data collection was related to the processes and machineries shared with products under the responsibility of the other two divisions located in the site under study. An example is the case of a bottling line that can be used to bottle both water and soft drinks. In this case, an allocation procedure was applied considering the volumes bottled in each bottling line.

Another challenge was related to the so-called "other activities" of the organization under study that cannot be directly referred to a single product and division. These refer, for example, to ancillary processes that serve different divisions (e.g., energy consumption and wastewater treatment). In these cases, specific allocation coefficients based on the company's historical data were developed and used. Data of upstream processes were collected starting from the identification of all relevant suppliers. Specific data collection among this supplier was performed. Downstream processes were modeled considering primary data on distribution and national statistics on use and end-of-life processes.

Data were also collected at the organizational level by adopting a top-down approach. These data were used to verify the energy and mass balance of the unit processes considered in the boundaries. Care should be taken when managing data aggregated at the organizational level; for instance, these data can be managed and registered by the company referring to either purchases (e.g., total quantity of PET grains acquired in the reference period) or effective use (e.g., total quantity of PET grains effectively used in production during the reference period).

5.4 Impact Assessment

Table 3 reports the results of the impact-assessment phase for the overall division under study and for the two product lines considered (PET- and glass-bottled water) with reference to the year 2013 (Fig. 5). These results are to be used as a baseline for the performance tracking of environmental impacts.

5.5 Interpretation of Results

The application of OLCA in this specific case study allowed aggregating the results at a different level of analysis (Fig. 6), thus allowing the identification of environmental hot spots and potential improvement strategies.

Table 3 Product portfolio of the division under study

Impact category	Method	Unit	Total potential impact	PET-bottled water	Glass-bottled water
Water scarcity	Pfister et al. (2009)	m^3	8.15E+05	7.80E+05	3.52E+04
Aquatic ecotoxicity	Impact 2002+ (Joliet et al. 2005)	kg TEG water	2.00E+10	1.80E+10	2.00E+09
Aquatic acidification		kg SO_2 eq	1.05E+06	8.84E+05	1.66E+05
Aquatic eutrophication		kg PO_4 P-lim	3.71E+04	3.26E+04	4.43E+03
Climate change	Recipe 2008 (Goedkoop et al. 2013)	kg CO_2 eq	2.31E+08	2.07E+08	2.37E+07
Ozone depletion		kg CFC-11 eq	2.13E+01	1.82E+01	3.14E+00
Terrestrial acidification		kg SO_2 eq	9.49E+05	7.98E+05	1.51E+05
Terrestrial ecotoxicity		kg 1,4-DB eq	1.68E+04	1.31E+04	3.69E+03
Human toxicity		kg 1,4-DB eq	5.19E+07	4.58E+07	6.10E+06
Photochemical oxidant formation		kg NMVOC	9.28E+05	8.03E+05	1.25E+05
Particulate matter formation		kg PM_{10} eq	3.34E+05	2.83E+05	5.13E+04
Ionizing radiation		kBq U235 eq	3.57E+07	3.17E+07	4.01E+06
Metal depletion		kg Fe eq	9.60E+06	8.68E+06	9.25E+05
Water depletion		m^3	6.44E+08	5.82E+08	6.14E+07
Fossil depletion		kg oil eq	8.59E+07	7.79E+07	7.94E+06

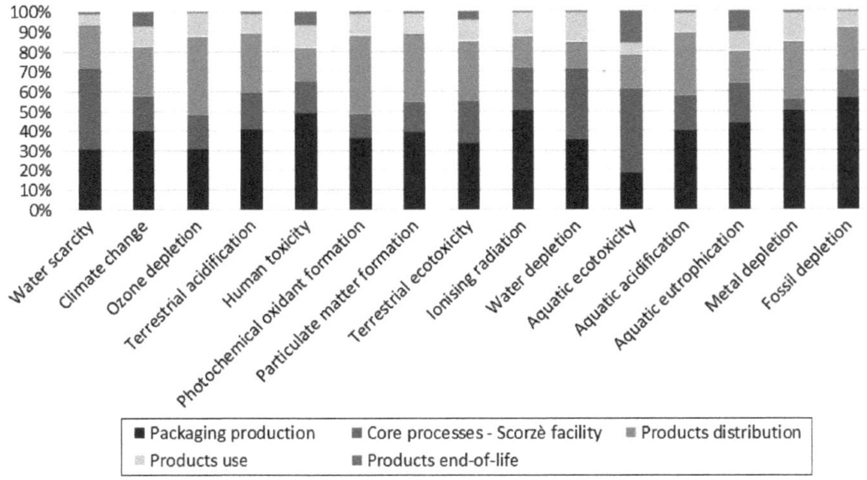

Fig. 5 Eco-profile at organizational level for PET-bottled water

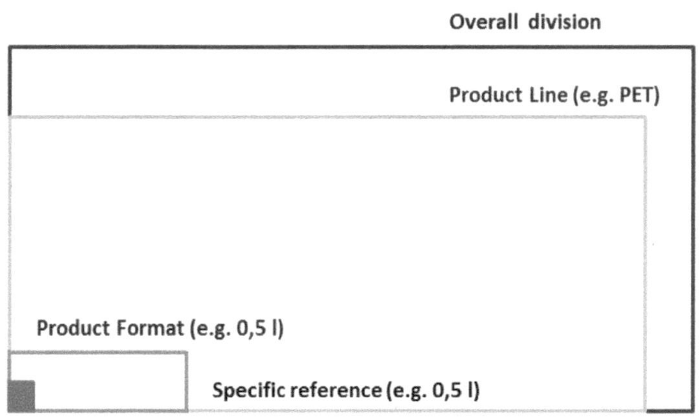

Fig. 6 Different level of analysis of the OLCA results

Table 2 shows that PET-bottled water has greater impacts in each of the impact categories considered. This is justified by the different quantities of water bottled with the two types of containers (980.211 m^3 in PET containers vs. 48.785 m^3 in glass containers) as well as their specific environmental impacts. Figure 7 shows the impacts per life-cycle stages in the category of climate change for the two product lines (PWT bottles and glass bottle). In both cases, the packaging is the responsible for most of the impacts on climate change. Looking at other impact categories, PET bottles contribute an average of 80 % of impacts and glass bottles an average of 55 % of impacts. Figure 8 shows the comparative analysis of

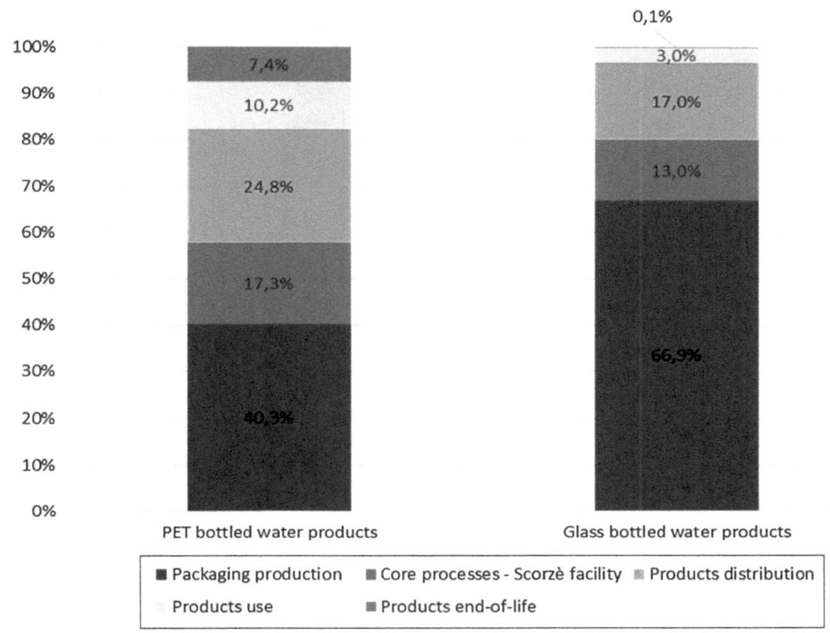

Fig. 7 Climate-change impacts of water bottled in PET versus glass containers

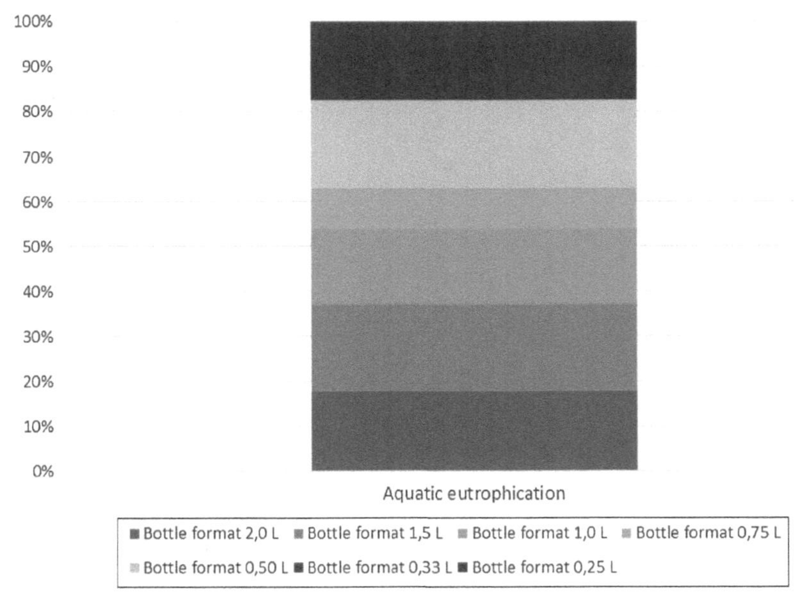

Fig. 8 Contribution of the different PET formats to overall PET impacts in the category of water eutrophication

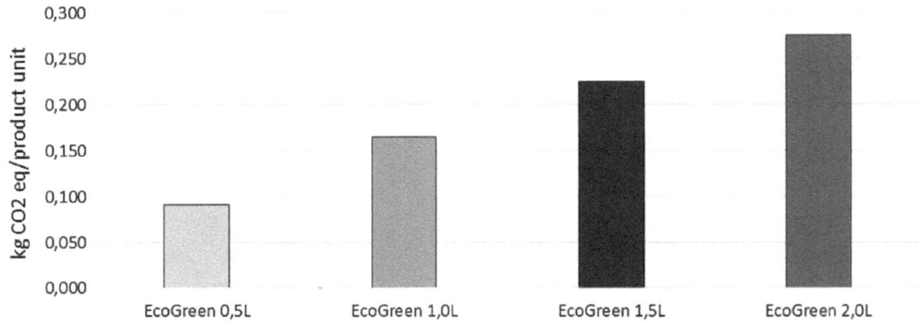

Fig. 9 Climate-change impacts of different product references

the impacts on eutrophication by different PET formats. These results depend on the quantity of water bottled in the different formats. OLCA can also be applied to study the impacts of single product reference. Figure 9 shows the results of impacts on climate change of a single product unit (branded San Benedetto Ecogreen) and allows comparison of their different environmental performances. Figure 10 represents an aggregation of results for all of the different PET bottles references, which has a relative contribution of >1 % to overall climate-change impacts. The representation allows understanding of which references the company should focus on to reduce climate-change impacts. The same representation can be performed for each of the impact categories selected in the goal and scope definition.

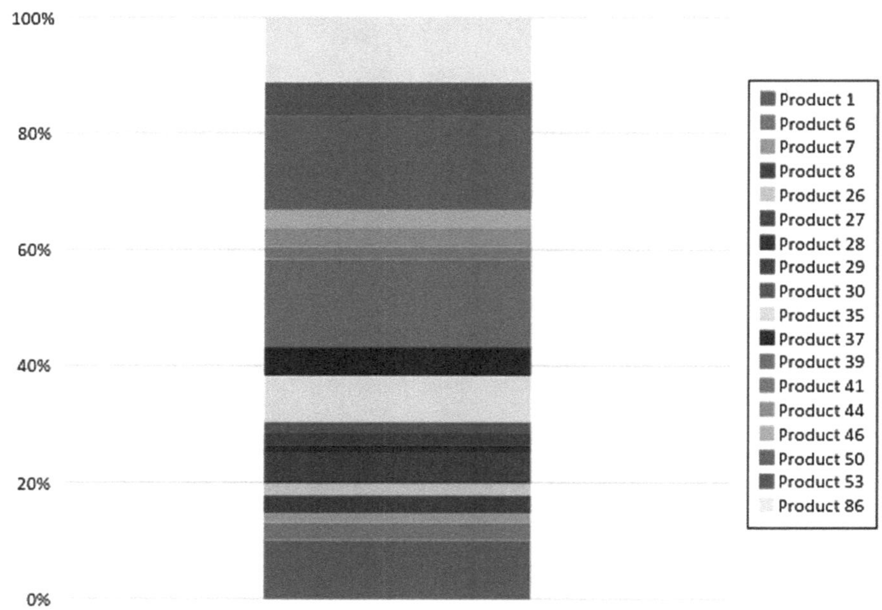

Fig. 10 Climate-change impacts of different product references

Life-cycle impacts of packaging are the processes where the company should focus to reduce its environmental burdens. This can be performed, for example, through container weight reductions and increased recycled material content along with improvements in the end-of-life operations.

6 Conclusions

OLCA is an emerging tool in the context of life cycle-assessment methodologies. Even if applications are limited due to the complexity of assessment and the novelty of the approach, its application can benefit companies from many perspectives. In the presented case study, the organization adopting OLCA was able to extend its capacity of control of overall impacts of its activities to the downstream and upstream processes using a life-cycle perspective.

Focusing on the results of the specific application presented in chapter "Environmental Impacts of Packaging Materials", future developments will be the extension of the model to all the other divisions of the organization. This model will also be tested to verify its capability to understand the mix of product references to be improved in order to minimize the overall environmental impacts. The model will also be periodically implemented for performance-tracking of the division included in the study.

Focusing on the experiences and results of the case studies presented in the literature, future developments can be identified to foster the application of OLCA in other contexts.

Specific procedures should be studied to simplify data collection; this could be achieved according to the potential integration of OLCA with EMS processes; moreover, the effects of data aggregation (purchase vs. consumption) on the final results should be investigated. Focusing on the potential uses of OLCA, it would be interesting to study how its application can support a company in the minimization of environmental impacts through delocalization of production facilities in reference markets.

References

Bayart JB, Worbe S, Grimaud J, Aoustin E (2013) The water impact index: a simplified single-indicator approach for water footprinting. Int J Life Cycle Assess 19:1336–1344

Boulay AM, Bulle C, Bayart JB, Deschênes L, Margni M (2011) Regional characterization of freshwater use in LCA: modeling direct impacts on human health. Environ Sci Technol 45:8948–8957

CDP (2014a) Carbon Disclosure Project—climate change program guidance website. https://www.cdp.net/enUS/Pages/guidance-climatechange.aspx

CDP (2014b) Carbon Disclosure Project—water program guidance website. https://www.cdp.net/en-US/Pages/guidance-water.aspx

DEFRA (2013) Environmental reporting guidelines: including mandatory greenhouse reporting guidance. Department for Environment, Food and Rural Affairs https://www.gov.uk/government/publications/environmental-reporting-guidelinesincluding-mandatory-greenhouse-gas-emissions-reportingguidance
EC (2009) Regulation (EC) No 1221/2009 of the European Parliament and of the Council of 25 November 2009 on the voluntary participation by organisations in a Community eco-management and audit scheme (EMAS), repealing Regulation (EC) No 761/2001 and Commission Decisions 2001/681/EC and 2006/193/EC
EU (2013) Recommendation of the Commission of the European Union on the use of common methods to measure and communicate the life cycle environmental performance of products and organizations. 9th of April 2013
EUROSTAT (2010) Environmental statistics and accounts in Europe. ISBN 978-92-79-15701-1. http://dx.doi.org/10.2785/48676
Finkbeiner M (2014) Product environmental footprint—breakthrough or breakdown for policy implementation of life cycle assessment? Int J Life Cycle Assess 19:266–271
Galatola M, Pant R (2014) Reply to the editorial "Product environmental footprint—breakthrough or breakdown for policy implementation of life cycle assessment?" written by Prof. Finkbeiner (Int J Life Cycle Assess 19:266–271). Int J Life Cycle Assess 19:1356–1360
Henningsson S, Hyde L, Smith A, Campbell M (2004) The value of resource efficiency in the food industry: a waste minimization project in East Anglia, UK. J Clean Prod 12:505–512
Hyde K, Smith A, Smith M, Henningsson S (2001) The challenge of waste minimization in the food and drink industry: a demonstration project in East Anglia, UK. Jon Clean Prod 9:57–64
Goedkoop M, Heijungs R, Huijbregts M, De Schryver A, Struje J, van Zelm R (2013) ReCiPe 2008. A life cycle impact assessment method which comprises harmonized category indicators at the midpoint and then endpoint level. First edition, Report I: Characterization
ISO (International Organization for Standardization) (2006a) ISO 14040: environmental management—life cycle assessment—principles and framework. Switzerland, Geneva
ISO (International Organization for Standardization) (2006b) ISO 14040: Environmental management—Life cycle assessment—Requirements and guidelines. Switzerland, Geneva
ISO (International Organization for Standardization) (2006c) ISO 14040: environmental management—Type III environmental declarations—Principles and procedures. Switzerland, Geneva
ISO (International Organization for Standardization) (2006d) ISO14064-1: Greenhouse gases—Part 1: specification with guidance at the organization level for quantification and reporting of greenhouse gas emissions and removals. Switzerland, Geneva
ISO (International Organization for Standardization) (2013) ISO/TR 14069 Greenhouse gases—Quantification and reporting of greenhouse gas emissions for organizations—Guidance for the application of ISO 14064-1. Switzerland, Geneva
ISO (International Organization for Standardization) (2014) ISO 14046 Water Footprint, Requirements and guidelines. Switzerland, Geneve
ISO (International Organization for Standardization) (2015a) ISO 14001 Environmental management systems—Requirements with guidance for use. Switzerland, Geneve
ISO (International Organization for Standardization) (2015b) ISO/TS 14072 Environmental management systems—Requirements with guidance for use. Switzerland, Geneve
Jolliet O, Margni M, Charles R, Humbert S, Payet J, Rebitzer G, Rosenbaum R (2005) IMPACT 2002+: A new life cycle impact assessment methodology. Int J LCA 8(6):324–330
Kounina A, Margni M, Bayart J-B, Boulay AM, Berger M, Bulle C, Frischknecht R, Koehler A, Milà i Canals L, Motoshita M, Núñez M, Peters G, Pfister S, Ridoutt B, Van Zelm R, Francesca F, Humbert S (2013) Review of methods addressing freshwater use in life cycle inventory and impact assessment. Int J Life Cycle Assess 18:707–721
Hellweg S, Milà i Canals L (2014) Emerging approaches, challenges and opportunities in life cycle assessment. Science 344:1109–1113
Manfredi M, Vignali G (2015) Comparative life cycle assessment of hot filling and aseptic packaging systems used for beverages. J Food Eng 147:39–48

Manzardo A, Ren J, Mazzi A Piantella A, Scipioni A (2014) Integration of water footprint accounting and costs for optimal chemical pulp supply mix in paper industry. J Clean Prod 72:167–173
Martínez-Blanco J, Inaba A, Finkbeiner M (2015) Scoping organizational LCA—challenges and solutions. Int J Life Cycle Assess 20:829–841
Notarnicola B, Hayashi K, Curran MA, Huisingh D (2012) Progress in working towards a more sustainable agri-food industry. J Clean Prod 28:1–8
Pfister S, Koehler A, Hellweg S (2009) Assessing the environmental impacts of freshwater consumption in LCA. Environ Sci Tech 43(11):4098–4104
Pelletier N, Allacker K, Pant R, Manfredi S (2014) The European Commission Organisation Environmental Footprint method: comparison with other methods, and rationales for key requirements. Int J Life Cycle Assess 19:387–404
Roy P, Nei D, Orikasa T, Xu Q, Okadome H, Nakamura N, Shiina T (2009) A review of life cycle assessment (LCA) on some food products. J Food Eng 90:1–10
Scipioni A, Mastrobuono M, Mazzi A, Manzardo A (2010) Voluntary GHG management using a life-cycle assessment approach. A case study. J Clean Prod 18(4)
Scipioni A, Manzardo A, Mazzi A, Mastrobuono M (2012) Voluntary GHG management using a life-cycle assessment approach. A case study. J Clean Prod 36:94–101
UNEP/SETAC (2015) Guidance on organizational life cycle assessment. Life-cycle Initiative, United Nations Environment Programme and Society for Environmental Toxicology and Chemistry, Paris, France. http://www.lifecycleinitiative.org/resources/reports/
UNFCC (United Nations Framework Convention on Climate Change) (1997) Kyoto protocol to the United Nations Framework Convention on climate change. In: Conference of the Parties to the United Nations Framework Convention on Climate Change. FCCC/CP/1997/7/Add.1
Weidema BP, Bauer C, Hischier R, Mutel C, Nemecek T, Reinhard J, Vadenbo CO, Wernet G (2013) Overview and methodology. Data quality guideline for the ecoinvent database version 3. Ecoinvent Report 1(v3). St. Gallen: The Ecoinvent Centre
WRI (World Resource Insititute) and WBCSD (World Business Council for Sustainable Development) (2001) GHG Protocol Corporate Accounting and Reporting Standard (GHG Protocol Corporate Standard). Earthprint Limited, USA, p 2001
WPO (World Packaging Organisation) (2008) Market statistics and future trends in global packaging. World Packaging Organisation/PIRA International Ltd a.m. pp 1–44

Potentials of Fibrous and Nonfibrous Materials in Biodegradable Packaging

Kartick K. Samanta, S. Basak and S.K. Chattopadhyay

Abstract Packaging is a, essential requirement for fruits, vegetables, agricultural crops, food products, and other commodities to provide the requisite protection from physical damage, contamination, deterioration; to increase shelf life; and facilitate need-based supply from the producer to the consumer. The packaging material should be physically and mechanically strong and should not add any foul odor to the packed product. In the past, for packaging of the above-mentioned products as well as various industrial goods has been made of traditional to advanced materials such as metal and glass; ordinary, coated, and laminated paper; corrugated paper box; gunny sack; textile bag; bamboo slit; wooden box; biodegradable film; nonbiodegradable plastic/film; composite; and nanocopmosite/biocomposite, all of which have been widely used. During the past 50 years, synthetic polymers have been found to steadily replace traditional packaging materials because of their advantages of low cost, low density, inertness, resistance to microbial growth, thermoplasticity, and transparency. However, their usage currently is being partially restricted because they are not totally recyclable and/or biodegradable and thus lead to serious environmental problem. This has resulted in the development of biodegradable polymers/films such as starch, polylactic acid, protein-based film, poly-beta-hydroxyalkanoates (PHB), etc. It has been possible to enhance physico-mechanical and functional properties of such polymers by incorporating organic and inorganic nanoparticles such as silver, titanium, chitosan, cellulose, clay, starch, silica, and zein. Similarly, traditional

K.K. Samanta (✉)
Chemical and Bio-Chemical Processing Division, ICAR-National Institute of Research on Jute and Allied Fibre Technology, Kolkata 700040, India
e-mail: karticksamanta@gmail.com

S. Basak · S.K. Chattopadhyay
ICAR-Central Institute for Research on Cotton Technology, Mumbai 400019, India

S.S. Muthu (ed.), *Environmental Footprints of Packaging*,
Environmental Footprints and Eco-design of Products and Processes,
DOI 10.1007/978-981-287-913-4_4

to coated/laminated paper/paper board, jute fabric, and the corrugated fibre board have been utilized for conventional to high-end packaging.

Keywords Biodegradable packaging · Fibrous · Jute · Nanocopmosite · Paper box

1 Introduction

Packaging is essential for fruits, vegetables, crops, and other commodities, including industrial products, to provide protection from physical damage, contamination, and deterioration as well as to increase shelf life. Packaging also ensures optimum distribution and storage costs, consumer convenience, and preservation of product quality and taste and facilitates need-based supply of the packaged goods from producer to consumer. Often such packaging, traditional to smart materials—such as metal and glass, plain, coated, and laminated paper, corrugated paper box, fabric, gunny sack, bamboo slit, wooden box, biodegradable film, nonbiodegradable plastic/film, composite, and nanocopmosite/biocomposite materials—are widely used. Packaging materials should possess strong physical and mechanical resistance properties to the nonthermal process (Galic et al. 2011). Nonthermal processes do not utilize increased temperature to inactivate decomposition of microorganisms and enzymes. All of the packaging materials can be broadly categorized into primary, secondary, and tertiary packaging. India is the second largest producer of fruits and vegetables next to China.[1,2] The country is also gifted with an abundance of different types of fruits and vegetables. Still, consumers are bereft of good-quality fruits and vegetables due to damage and/or spoilage of approximately 20–30 % such products during transportation (see Footnotes 1 and 2).[3,4] Therefore, a good package must withstand the stress and strain of long-distance transportation, multiple handling, and change in climatic conditions at different storage places. Several technological advancements have taken place during the past 20 years in the packaging of food products with the evolution of the society and its lifestyles. Indications are strong and clear from recent research and developments that food packaging will continue to evolve in response to the increased consumer needs and futuristic demands. The proper selection and optimization of

[1]http://theglobaljournals.com/paripex/file.php?val=June_2013_1371303434_72a9a_08.pdf, dated 30-06-2015.

[2]http://www.nird.org.in/NIRD_Docs/rs2013/RS%2091.pdf.

[3]http://www.itfnet.org/gfruit/Slides/Session%202/Marketing%20of%20Fruits%20in%20India%20-%20Present%20Practice%20and%20Future%20needs.pdf, dated 30-06-2015.

[4]http://www.business-standard.com/article/economy-policy/reducing-wastage-of-fruits-vegetables-is-the-key-focus-since-it-would-help-to-address-inflation-union-food-minister-box-attached-114091800774_1.html, dated 30-06-2015.

packaging materials are important to food manufacturers due to the associated aspects of economics, marketing, logistics, distribution, environmental impact, and consumer demands. Today, the packaging industry relies strongly on the use of petroleum-derived plastic materials, which is raising concerns on both environmental and economic impacts (Lavoine et al. 2014). In addition to traditional packaging materials, new research is focused on functional packaging materials, such as antibacterial and conductive, to improve product quality and keep it free from microbial spoilage.

The production of plastic materials for packaging application has seen a dramatic increase in the last two decades, and synthetic polymers have also been steadily replacing traditional packaging materials, such as paper, glass, metals, etc., during the past 50 years mainly because of their low cost, low density, inertness, ease of availability, resistance to corrosion, softness, transparency, and possessing the desirable physical (e.g., barrier and optical) and mechanical properties (Siracusa et al. 2008). Most of the plastics are made of chemicals that are derived from crude petroleum oil. However, their use is now being restricted because they are not totally recyclable and/or biodegradable and thus pose a serious threat to the environment. Similar to synthetic and nonbiodegradable polymer-based packaging materials, textile (fibrous)-based packing materials also play a crucial role in packaging applications. Different textile structures, especially designed for the packaging of food grains, sugar, rice, cement, other commodities, and industrial goods, are known as "Packtech" in technical textile parlance. The textile structures include use of both natural fibres, such as jute and cotton, and synthetic petroleum based fibres, such as polyester, polyethylene, polypropylene, etc. Uncoated and coated/laminated textiles, as well as paper-/pulp-based single to multilayer bags/structures, are also used as a shopping bags, food packets, and in the packaging of agricultural commodities due to their advantages of biodegradability, structural flexibility, and cost-effectiveness. A large quantity of jute hessian and sacking bags are also used as packaging materials in India because the country is the largest producer of jute fibre globally and the second largest exporter of jute goods, which ultimately supports the livelihood of 40 lakh farm families.[5]

The emerging nanotechnology has also been explored in the food and packaging sectors to enhance physical, mechanical, and functional properties of paper, film, and composite packaging materials (Youssef et al. 2013). Inorganic (e.g., silver and titanium) and organic nanoparticles (e.g., chitosan, cellulose, clay, starch, silica, and zein protein) with particle sizes in the range of 10–500 nm have been synthesized and incorporated into various biopolymers as fillers or coating materials to enhance the barrier, mechanical, and functional properties of packaging materials. The reinforcement of biopolymers using natural fillers, such as fibre, fibril, and organic nanoparticles, has attracted consideration because it is applied in an environment friendly manner for the development of the advanced materials. Such developed products are also environmentally consistent because both the

[5]http://www.wbidc.com/images/pdf/annual_report/annual_report-09-10/Jute-Industry.pdf, dated 22-05-2015.

matrix and the filler are produced/derived from a renewable source, such as agricultural residues (e.g., parts of plants), or natural resources.

2 Importance of Packaging Material

Packaging is connected substantially and intimately to our everyday life, and its use has steady increased over time. With the development of the society and due to the availability of diversified food, e.g., fast food, junk food, and functional food, there has been an increasing requirement for traditional to high-end packaging materials, thus accelerating the development of new food-packaging materials. It is also expected that the packaging material should be physically and mechanically strong, should be free from contamination, and should not add any foul odor to the packed product. Therefore, a food product is packaged with the aim of storage, preservation, and protection for long-term use.[6] These are the three basic attributes demanded from food-packaging technology that must be perfected for better quality and handling of foods. A wide range of materials (e.g., metal, glass, wood, bamboo slit, paper- or pulp-based materials, fabric, and plastics) or combinations of materials (e.g., composites) are used for the packaging of foods and other commodities. The per-capita consumption of plastics in the United States, for example, is approximately 150 kg, in Europe approximately 20 kg, and in India approximately 5 kg (Nayak and Swain 2002). In developed countries, such as the United Kingdom, the proportion of food that is unfit for consumption before it reaches the consumer is 2 %, whereas in developing countries, where packaging is not as widespread, this loss can be in excess of 40 % (Davis and Song 2006). Almost all packed and traded consumer goods should fulfill at least one of the below-motioned functions in day-to-day life (Galic et al. 2011; Davis and Song 2006).[7]

- provide protection from physical damage, contamination, and deterioration
- offer sale appeal
- ensure product identity
- provide information about the product
- optimize distribution and storage costs
- provide consumer convenience and safety

Packaging materials play a major role in ensuring microbiological food safety by acting as a physical barrier preventing external contaminants coming into contact with the food. Additionally, they also fulfill the important function of protecting the packaged food from light, oxygen, and humidity, thus enhancing the shelf life of the product (Feichtinger et al. 2015). Packaging also plays a critical role in the

[6]http://en.wikipedia.org/wiki/Packaging_and_labeling, dated 22-05-2015.

[7]http://en.wikipedia.org/wiki/Food_packaging, dated 22-05-2015.

postharvest handling and distribution of fresh and processed foods and other biomaterials (Pathare and Opara 2014). Similarly, in the long and complicated journey of fresh horticultural produce from producer to consumer, packaging is very important. Paper and cloth are flexible and lightweight, and generate less waste to discard in terms of packaging materials. Indeed, glass and metals have been used for packaging high-value products because they are corrosion resistant and stronger. On the other hand, the well-explored polymer (plastics) materials are extensively used for high-value packaging with an annual world production of approximately 200 MT and an average per-capita consumption of 100 kg (Mahalik and Nambiar 2010). This is due to their desirable properties such as tear resistance, tensile strength, excellent barrier to oxygen, thermal-seal ability, transparency, and softness; they are also inexpensive to produce (Mahalik and Nambiar 2010). All packaging materials can be broadly categorized into the following three groups: primary, secondary, and tertiary.

Primary packaging: Primary packaging usually remains in contact with the goods taken home by consumers. The most common types of materials used in this category are paper or pulp, glass, metals (aluminum and steels), and plastics. Paper- or pulp-based materials, such as wrapping paper, carton boxes, disposable cups and plates, bags, and envelopes, and corrugated cardboard, are used as both primary and, to some extent, secondary packaging.

Secondary packaging: Secondary packaging includes larger packaging, such as boxes, used to carry a number of primary packaged goods.

Tertiary packaging: Tertiary packaging refers to packaging, such as wooden pallets and plastic wrapping, used to assist in the transport of large quantities of goods.

Secondary and tertiary packaging materials are normally used in larger quantities and have less material variation; thus, recollecting and sorting them by wholesalers or retailers for recycling or reuse are much easier. Unlike secondary or tertiary packaging, primary packaging materials are not only more dispersed into households but also are mostly mixed, contaminated, and often damaged, and therefore they pose considerable challenges in recycling or reuse (Davis and Song 2006).

Currently a large number of petrochemical-based polymers, namely, PET, PP, PE, PS, and PA, are being used for the packaging of foods, crops, chemicals, fertilizers, and various industrial products owing to their low cost, light weight, inertness, transparency, and availability in large quantity. Because they are commonly derived from petroleum origin, they are nonbiodegradable and difficult to recycle or reuse due to their mixed levels of contamination and composition in addition to the presence of different polymer additives. A large number of traditional-to-smart fibrous and nonfibrous materials (e.g., metal, paper, coated and laminated paper, fabric, gunny sack, coated/laminated fabric, corrugated paper box, bamboo slit, wooden box, etc.) are widely used in the packaging (e.g., shopping bags, food packing, industrial products, fertilizer, cement, tea, etc.) of agricultural crops and commodities due to their biodegradability, their flexible to semi-flexible structure, and their cost-effectiveness. Also, in the recent years there has been a paradigm shift toward the development of biodegradable polymers and packaging materials, and some of the developments that are getting attention in this context are starch, polylactic acid, protein-based film, poly-beta-hydroxyalkanoates (PHB), etc. Their inherent physical, mechanical, and functional properties, and the incorporation of various micron- to nano-size fillers, will be discussed in detail in successive sections of this chapter.

3 Biodegradable Packaging Materials

The current global consumption of plastics is >200 million tonnes with an annual growth rate of approximately 5 %, which represents the largest field of application of crude oil (Siracusa et al. 2008). Until now, petrochemical-based plastics, such as polyethylene terephthalate (PET), polyvinylchloride (PVC), polyethylene (PE), polypropylene (PP), polystyrene (PS) and polyamide (PA), have been increasingly used as packaging materials because of their techno-economic-mechanical advantages of availability in a large quantity at a relatively lower cost, good tensile and tear strength, good barrier to oxygen, carbon dioxide, anhydride, and aroma compounds, heat sealability, thermoplasticity easiness in making flexible and semi-flexible bag/structure, and relatively inertness to the packaged product (Vigneshwaran et al. 2011). Plastics or synthetic polymers are the long-chain molecules that started to substitute for natural materials in almost each and every area of applications approximately half a century ago; currently plastics have become an indispensable part of our lives (Shah et al. 2008). With the passage of time, the stability and durability of plastics have been improved continuously; hence such materials are now considered synonymous for materials that are resistant to many environmental influences. The basic materials used for making plastics are extracted from oil, coal, and natural gas. However, they are neither totally biodegradable nor recyclable, thus causing adverse effects to the environment, especially soil and water (Mahalik and Nambiar 2010; Shah et al. 2008). Plastic packaging materials are also often contaminated by foodstuff and biological substances; therefore, their recycling is impracticable and, most of the time, economically inconvenient (Siracusa et al. 2008). As a result, a several thousand tons of such materials are land-filled, which increases the environmental problem day by day. Due to the adverse effect of such fossil fuel-based polymer (99 % of plastics are made from fossil fuel) in packaging, there has been a paradigm shift in recent years toward the development of biodegradable polymers and packaging materials to address such environmental issues (Mahalik and Nambiar 2010; Azeredo 2009). Biodegradation is the process by which carbon-containing chemical substrates are decomposed in the presence of enzymes secreted by living organisms.

More recently the development of biodegradable packaging materials from renewable natural resources has received widespread government support in European Union countries. The field of application of biodegradable polymers in food-contact articles includes disposable cutlery, drinking cups, salad cups, plates,

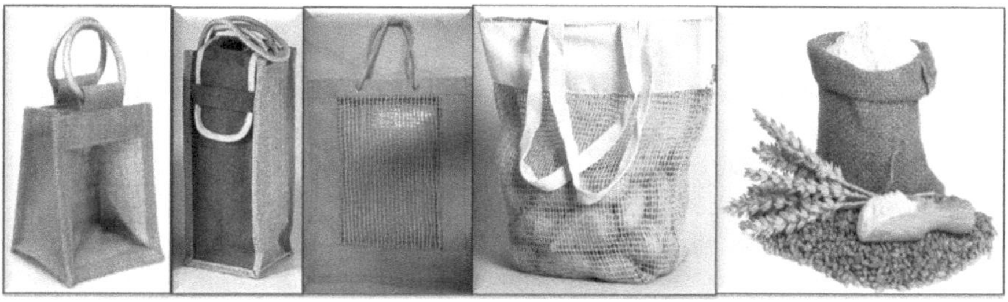

Fig. 1 Different packaging materials made of jute and cotton

overwrap and lamination film, straws, stirrers, lids, cups and plates, and containers for food dispensed at delicatessens and fast-food establishments. By biological degradation, such biodegradable polymers produce water, carbon dioxide, and inorganic compounds but no toxic residues. According to the European Bioplastics Norm, biopolymers made of renewable resources must be biodegradable and especially compostable so they can act as fertilizers and soil conditioners at the end of their life (Siracusa et al. 2008). Bioplastics, such as plastics, also present a large spectrum of applications such as collection bags for compost, agricultural foils, horticultures, nursery products, toys, fibres, textiles, etc.[8] Some of the explored biodegradable polymers suitable for packaging application are starch, poly(lactic acid) (PLA), cellulose, zein protein, poly-beta-hydroxyalkanoates (PHB), polyhydroxy-*co*-3-butyrate-*co*-3-valerate (PHBV), and others. One of the most promising biopolymer is PLA obtained from the controlled polymerization of lactic acid monomer, which is obtained from the fermentation of sugar feedstock, corn, etc., which are in turn obtained from renewable resources; thus, they are readily biodegradable. PLA is a versatile recyclable and compostable polymer with high transparency, high molecular weight, and good processability and water-insolubility. Currently it is used in food-packaging applications only in the cases of products with short shelf lives. Such properties have also been observed in starch for packaging applications. Similar to recent development in biodegradable polymers and films, natural fibres-based woven and nonwoven fibrous structures have also been used in biodegradable packaging for a long time. For example, cellulosic cotton and lingo-cellulosic jute fibres have extensively been used for the packaging of agricultural crops, sugar, fertilizer, and shopping bags as shown in Fig. 1.

4 Natural Fibres-Based Packaging Material

One of the important uses of textiles is the manufacturing of various bags and sacks not only from traditional cotton, flax, and jute fibres but also from the synthetics such as polypropylene.[9] Different textile materials that are especially used

---

[8]www.european-bioplastics.org, dated 22-05-2015.

[9]http://www.technotexindia.in/packaging-textiles.html, dated 22-05-2015.

for the packaging of various commodities fall under the group of "Packtech" under the umbrella defining the technical textiles or functional textiles. Products covered under Packtech range from polymer-based bags used for industrial packing to jute-based sacks used for packaging of food grains and tea. These kinds of packaging materials (excluding jute) are also called "flexible packaging materials." The ability to reuse these containers in many applications in place of disposable bags and sacks further supports their wider use. Some other products under Packtech include polyolefin-woven sacks, leno bags, wrapping fabric, jute hessian and sacks (including food-grade jute bags), soft-luggage products, tea bags (filter paper), and others.[10]

4.1 Jute Textile-Based Packaging Material

In India as per the government norms as published in The Gazette of India under section 376, a minimum 90 % of food grains (after providing for upfront exemption of 3.5 lakh bales) and 20 % of sugar of the total production of the country are to be packed in jute fibre-based hessian and sacking bags (The Gazette of India, Extraordinary, Part II-Section 3-Sub-section (II), Ministry of Textile-Order dated 13th Feb 2015). Natural fibre, such as jute, is most suitable for the packaging of sugar and other agricultural food grains owing to its advantage of low cost, biodegradability, eco-friendliness, produced from renewable sources, yet it is capable of satisfying the standards for safe packaging compared with synthetic HDPE and PP bags. Jute packaging material means jute fibre, jute yarn, jute twine, jute sacking cloth, hessian cloth, jute bags, or any other packaging material that contains jute fibre not less than 75 % by weight. The role of the jute industry in the Indian economy is very important because it is the major industry in the eastern part of India, particularly in the state of West Bengal (see Footnote 5). Jute, an important cash crop, is intercropped before paddy transplantation in most parts of the country. India is the largest producer of jute globally and the second largest exporter of jute goods, which ultimately supports 40 lakh farm families' livelihood. Jute fibre is mostly used for the packaging of agricultural crops, rice, sugar, tea, potato, etc. However, with the development of petroleum-based low cost and lightweight synthetic bags, such as high-density polyethylene (HDPE) and polypropylene (PP), jute bags have slowly been replaced by such bags. As a result, the jute industry has been phased out from Europe, America, and the far East, and today it survives only in the Indian subcontinent and, to some extent, in Brazil and China. Raw jute in the form of bales is processed in the jute industry to produce hessian, sacking, jute yarn, bags, and other useful products. Raw jute bales weighing approximately 150 or 180 kg with or without a top portion being cut generally come to the factory and are assorted according to their suitable end-use, such as hessian weft,

[10]http://textilelearner.blogspot.in/2013/01/packtech-textile-packaging-material.html, dated 22-05-2015.

sacking wrap, sacking weft, etc. Before spinning, the fibres are softened in softener or a spreader machine to lubricate and/or to soften the bark and the gummy portion of the raw jute fibre by application of an oil–water emulsion.

A large amount of jute-based hessian and sacking bags is procured by the Government of India for the packaging of agricultural food grains. Therefore, several standards have been specified for different end applications by the government. A numbers of bags or lap of fabrics, also called "cuts," whichever is applicable, are packed with the help of a packsheet and bailing hoops to form a compact package called a "bale." The standards have been formulated to mitigate the requirements of good packaging, which includes the degree of thrust-withstanding capacity, seam strength of the bags, and the prevention of leakage of packed material through the bag. A "lot" consisting of numbers of bales must fulfill all the requirements and criteria for conformity as specified in the standards. The requirement consists of standards and tolerance of length, width, ends/dm and picks/dm, weight, and moisture regain of the bag. Apart from this, acceptance criteria for strength and manufacturing defects of the fabric are also specified in the respective standard. For example, the Indian Standard (2nd Revision) for Textiles–Jute bags for packing of 50 kg of food grains was adopted by the Bureau of Indian Standards, and the draft finalized by the Jute and Jute Products Sectional Committee has also been approved by the Textile Division Council. The bags shall be made from a single piece of double warp, 2/1-twill weave jute sacking of uniform construction having a nominal mass of 579 g/m^2 with the warp running along the length of the bag (Indian Standard (IS) 2003). There shall be a single blue stripe or stripes woven along the length of the bag or the bag shall be without stripe as agreed between the buyer and the seller. The constructional parameters of such sacking bags are indicated in Table 1. This kind of bags is mostly suitable for the packing of wheat, rice, and similar coarse grains. However, for packing of other materials, the buyer and the seller may agree to the dimensions other than those specified in this standard. The sides of the bags shall be sewn with overhead or herakle stitches on the selvedge through two layers of the sacking, and the number of stitches per decimeter shall be 10 ± 1. In the defect test, a bag shall be termed as defective if it contains two or more major defects, such as the GAW (>1.5 cm), multiple broken/missing warp end (single, >25 cm long), multiple broken weft pick (two or more continuous regardless of length), float (>2 cm^2), gap stitching (stitches missing >1.5 cm), and corner gap (>1.5 cm).

Similar to the above-mentioned criteria, IS 15138: 2010 on jute bags for packing 50 kg of sugar is one of the most widely implemented standards because it covers comprehensive specifications of the raw material and their classification, dimensional requirements, physical and chemical characteristics, mechanical properties, sampling criteria, test requirements, and acceptance criteria as indicated in Table 2 (Indian Standard (IS) 2010). A jute sack is woven on conventional shuttle looms as well as modern rapier looms and it is usually available in plain and twill woven form. Jute sack, commonly known as "heavy goods," is loosely woven, weighs from 12 to 20 oz per yard, and comes in different widths depending on the kind of goods to be packed. They are commonly utilized for packaging

Table 1 Specification of jute bags for 50 kg of food grains packaging (Indian Standard (IS) 2003)

Sl. no.	Characteristics	Requirement		Tolerance
		Type A	Type B	
1	Bag dimension			
	(i) Outside length, cm	94	94	+4 cm, −0
	(ii) Outside width, cm	57	57	+4 cm, −0
2	Ends/dm	76	46	+4, −3
3	Picks/dm	28	50 (2 × 25)	+2, −2
4	Corrected mass/bag, g	665	665	+8 %, −6 %
5	Average breaking strength of fabric (Ravelled-strip method: 10 cm × 20 cm), Min., N (kgf)			
	(i) Warp way	1570 (160)	1570 (160)	–
	(ii) Weft way	1420 (145)	1420 (145)	–
6	Average seam strength (5 cm × 20 cm ravelled strip), Min. N (kgf)	490 (50)	490 (50)	–
7	Moisture regain, percentage, Max.	22	22	–
8	Oil content on dry de-oiled material basis, percentage, Max.	3	3	–

Type A Single warp, double weft woven on modern shuttleless loom
Type B Double warp, single weft woven on conventional shuttle loom

Table 2 Specifications of jute bag for 50 kg of sugar packaging (Indian Standard (IS) 2010)

Sl. no.	Characteristics	Requirement		
		Type A [Tolerance[a]]	Type B [Tolerance[a]]	Type C [Tolerance[a]]
1	Bag dimension			
	(i) Outside length, cm	87.5 [+3]	91.5 [+3]	91.5 [+3]
	(ii) Outside width, cm	58.5 [+3]	56.0 [+3]	56.0 [+3]
2	Ends/dm	68 [+4, −2]	47 [±2]	47 [±2]
3	Picks/dm	31 [+2, −1]	55 [+2, −1]	47 [+2, −1]
4	Corrected mass/bag, g [Tolerance in %, Max.]	630 [+7.5, −6]	475 [+7.5, −2]	405 + 32 Liner [+7.5, −2]
5	Average breaking strength of sacking (Ravelled-strip method: 10 cm × 20 cm), Min., N (kgf)			
	(i) Warp way	1570 (160)	1470 (150)	1470 (150)
	(ii) Weft way	1420 (145)	1765 (180)	1420 (145)
6	Average seam strength (5 cm × 20 cm ravelled strip), Min. N (kgf)			
	(i) Warp way	–	490 (50)	490 (50)
	(ii) Weft way	440 (45)	685 (70)	490 (50)
7	Moisture regain, percentage, Max.	22	17	17

[a]Value in the parenthesis is the tolerance limit/percentage

of bulky articles weighing 50–100 kg such as sugar, wheat, tea, rice, etc. Type A bags shall be made from a single piece of 568 g/m^2 plain-weave construction double-warp jute fabric with the warp running along the length of the bag. Type B and C bags shall be made from hessian having a mass of 417 and 354 g/m^2, respectively. The cloth shall be without stripes or shall have stripes woven along the length of the bag as per the agreement between the buyer and the seller. This kind of packaging bag for sugar shall be specifically manufactured from raw jute of Indian origin. The sides of the type A bag shall have herakle safety stitches per the standard norm, and type B and C bags shall be sewn with herakle stitches on the selvedge through the two layers of jute, and the bottom row edge shall be folded inside to a depth of at least 3.8 cm and then stitched at the mouth.

A similar specification for the laminated jute bags for the packaging of milk powder has also been standardized per the Indian Standard IS 12626: 1989 (Indian Standard (IS) 1989). Here, there are two types of bags: The Type 1 bag is made out of hessian fabric laminated with kraft paper on the outside and plastic film/kraft paper on the inside, uses bitumen as the bonding agent, and has a liner of kraft paper or plastic film stitched along the bag on the side and at the bottom. Type 2 bags (with GSM 270 g/m^2) are made of hessian fabric laminated with kraft paper on the outside, use bitumen as a bonding agent, and have two liners of kraft paper or plastic film stitched along the bag on the side and the bottom. The bags are required to be manufactured from the laminated hessian fabric with stitching on the side and the bottom to keep the kraft paper on the outer side of the bag. In this regard, low-density polyethylene (LDPE), high-density polyethylene (HDPE), high-molecular high-density polyethylene (HMHDPE), or polypropylene (PP) film shall be used. In case of low-density polyethylene film, the areal density of 23 g/m^2 or a thickness of 25 μm, and for other type of plastic films a minimum mass of 11.5 g/m^2 or a thickness of 12.5 μm, are recommended. The liner for food-grade material, i.e., the LDPE, HDPE, HMHDPE, or PP loose liner, shall be of minimum mass of 69 g/m^2 or thickness of 75 μm. The detail specifications, including the physical and mechanical properties, are also reported in this standard (Indian Standard (IS) 1989). The jute fabric has also been recommended for packing of fertilizer per the Indian Standard IS: 7406 (Part 1)-1984 (Indian Standard, IS: 7406 (Part 1) 1984). Double-warp jute tarpaulin bags are used conventionally for the packaging of fertilizer. Such bags frequently undergo adverse climatic conditions and transport hazards from factory to the farmer's field. The size of the bags is specified in such a way so they hold 50 kg of fertilizer. Bags suitable for lower bulk-density fertilizer are approximately 99 cm × 61 cm with a tolerance of ±3 cm. The specifications of the laminated jute bags manufactured from 380 g/m^2 fabric and 68 cm × 39 cm tarpaulin fabric have been covered in Part 2 of the same standard (Indian Standard 1986). Similar to other specifications, IS 9685: 2002 describes the textiles suitable for the packaging of sand (e.g., sand bags) (Indian Standard, IS 9685 2002). The bags shall be made from one continuous piece of 229 g/m^2 hessian, and each piece may be folded widthwise or lengthwise, but the bag length shall be in the direction of the warp of the fabric. If there is a requirement of rot-proofs by the buyer, the sand bags shall be finished with copper naphthenate per standard IS 11662. The outer length and width of the bag should be approximately 84 cm × 36 cm with a tolerance of +3 cm, and the mass of the bags shall be 160 and 180 g for nonproofed and rot-proofed bags, respectively. As per the Indian Standard IS 12269: 2013, cement shall be packed in any of the following bags: (1) jute sacking conforming to IS 2580, (2) light-weight jute fabric conforming to IS 12154, (3) jute synthetic union bags conforming to IS 12174, (4) multiwall paper sacks conforming to IS 11761, and (5) HDPE/PP woven sacks conforming to IS 11652 (2013). Thee bags shall be prepared in such a way that the cement capacity per bag will be approximately 50 kg. However, the net quantity of cement per bag may also be 25, 10, 5, 2, or 1 kg subject to the acceptable tolerances and packed in such bags per the mutual agreement between the purchaser and the manufacturer.

As per IS 1943, an A-twill bag shall be made from a single piece of double-warp, 2/l twill-weave jute sacking of uniform construction and having a nominal mass of 750 g/m^2 with the warp running along the length of the bag as indicated in Table 3. There shall be three blue stripes or simple stripes along the length of the bag as per agreement between the buyer and the seller (IS 1943 1995). The sides of the bags shall be sewn with overhead or herakle stitches on the selvedge through two layers of sacking with 9 to 11 stitches/10 cm. A line of safety union stitch with the above stitch density shall also be provided at the inner edges of the overhead or herakle stitches. The bags should preferably be free from weaving and sewing defects, such as missing picks, holes, cuts, tears, float, crushed selvedges, spots, stains, gap stitches, loose ends, and frayed ends, which might affect the end performance of the bag as a packaging material. The bag shall be made of 750 g/m^2 areal density fabric with 102 ends and 35 picks/dm where the acceptable tolerance limits are ±6 and ±2, respectively. The outer dimension of the bag shall be 112 cm in length and 67.5 cm in width, with a total bag-weight

Table 3 Specification of an A-twill jute bag for sugar packaging (Indian Standard 1995)

Sl. no.	Characteristics	Type A [Tolerance[a]]
1	Bag dimension	
	(1) Outside length, cm	112 [+4, −0]
	(2) Outside width, cm	67.5 [+4, −0]
2	Ends/dm	102 [+6, −6]
3	Picks/dm	35 [+2, −2]
4	Corrected mass/bag, g [Tolerance in %, Max.]	1200 [+10, −7.5]
5	Average breaking strength of sacking (ravelled-strip method: 10 cm × 20 cm), Min., N (kgf)	
	(i) Warp way	2000 (204)
	(ii) Weft way	1765 (180)
6	Average seam strength (5 cm × 20 cm ravelled strip), Min. N (kgf)	657 (67)
7	Moisture regain, percentage, Max.	22

[a]Value in the parenthesis is the tolerance limit/percentage

of 1190 g. It is recommended that the product should also ensure a warp and weft way breaking load of 2000 and 1765 N, respectively, and a seam-breaking load of 657 N.

Similar to an A-twill jute bag, the standard for a B-twill jute bag describes the construction details along with other requirements of the bag for the packing of 100, 93, and 75 kg of food grains (Indian Standard 1993). The corresponding bag size will be 122 × 67.5 cm, 112 × 67.5 cm and 106.5 × 61 cm, respectively, and all the bags should have 76 and 31 ends and picks/dm, respectively. The weight of the bag for the overhead stitch shall be 1110, 1020, and 880 g, respectively. It may be noted here that the packing of 50 kg of food grains is covered under IS 12650: 1989. The bag shall be made from cloth conforming to IS 3667: 1993. It should be made from a single piece of cloth preferably with the warp running along the length of the bag. For marking purposes, a blue stripe of single or double warp shall be placed per the agreement between the buyer and the seller. Similarly, the bags should also be free from the defects as stated previously.

The announcement to exempt 60 % of the output of the sugar industry from jute packaging has provided a breather to the industry. Earlier, under the Jute Packaging Materials Act (1987), sugar manufacturers had to package the entire product in jute bags.[11] A jute bag costs approximately Rs 35 for a 50-kg bag, which translates into a packaging cost of Rs 0.70/kg of sugar. On the other hand, a similar capacity, high-density polyethylene (HDPE) bags costs much less, approximately Rs 15, and hence the packaging cost becomes Rs 0.30/kg of sugar. Therefore, a shift from a jute bag to HDPE bags would result in savings of Rs 0.40/kg for sugar packaging. Assuming that the total production is 25 million tonnes and that the entire product is packaged in HDPE bags, the total savings for the industry would stand at Rs 1000 crore. However, because presently the government has allowed HDPE bags for only $\leq$80 % of the produce, the industry's savings would stand approximately at Rs 800 crore.

4.2 Other Textile-Based Packaging

Textiles meant for packaging include all the textile-based materials for packing of industrial, agricultural, and other goods. The demand for packing material is directly proportional to economic growth, industrial production, and trade to distribute goods both locally and internationally.[12] As discussed previously, packaging has provided new opportunities in the emerging marketplace regarding the growing environmental need for reusable/recycled/biodegradable packages and containers and natural fibres-based products. Sacks and bags made of traditional jute, cotton, or natural fibers are gradually making their way into the market by

[11] http://www.business-standard.com/article/markets/packaging-shift-to-help-sugar-sector-save-rs-600-cr-a-year-112110800052_1.html, dated on 07-0502915.

[12] http://www.bch.in/packaging-textiles.html, dated 22-05-2015.

replacing those made from synthetic fibers. Those technical textiles, specially used in packaging and subsequent transportation applications, are called Packtech. Packtech includes not only heavy-weight, densely woven fabrics (bags and sacks for storage, flexible intermediate bulk carriers, and wrappings for textile bales and carpets) but also lightweight woven and nonwoven fabrics used for durable papers, tea bags, shopping bags, industrial product wrappings, woven strapping, light-weight mailbags, soft luggage, and coffee filters (see Footnote 12).[13]

5 Paper, Pulp, and Corrugated Boxes in Packaging

5.1 Paper/Paperboard and Paper Pulp in Packaging

The use of paper pulp in packaging has become more attractive compared with traditional materials, such as expanded polystyrene (EPS) foam, due to the advantages of low price of the recycled paper, low cost of production, and biodegradability (Gurav et al. 2003). It has been noticed that electronic equipment, such as computer monitors, could be better packed in a paper pulp packaging than the traditional expanded polystyrene packaging. Paper pulp is basically a composite materials made out of recyclable waste papers, such as newsprint or cardboard, that are made from naturally available materials consisting of wood fibres in a matrix of lignin and hemi-cellulose. In addition to the waste paper, the process requires only water and energy to produce the paper pulp. Hence, it helps in saving resources and becomes an "eco-comp" material. Paper/paperboard has great advantages compared with traditional plastic as packaging material in terms of cost and sustainability (Chen et al. 2013). The primary consumption of this kind of material is for producing various types of flexible, semi-rigid, or rigid packaging. Due to its overall advantages of cost-effectiveness, high flexibility, environmental friendliness, produced from renewable sources, and ease of recyclability, these types of packaging are used in the largest quantity as packaging materials throughout the world. The global paper-packaging market in 2011 was reported to be worth of 236 billion dollars USD (Chen et al. 2013). The typical composition of pulp is listed in Table 4. It has been reported that various jute-based raw materials, such as fibres, stick, mesta stick, whole plant, feshwa, root cuttings, caddis, etc., are a good source of cellulose and hence has been explored to produce paper-grade pulp and ultimately utilized to produce different grades of important papers and board. Similarly, an adhesive-bonded fabric of 110 g/m^2 has been developed to make light-weight carry bag.[14] Some of the bags produced from jute pulp/paper and light-weight jute fabric at ICAR—National Institute of Research on Jute and Allied Fibre Technology are shown in Fig. 2.

[13]http://en.wikipedia.org/wiki/Technical_textile, dated 22-05-2015.

[14]www.nirjaft.res.in, dated on 29-06-2015.

Table 4 Composition of pulp (Gurav et al. 2003)

Material	Structure	Approximate weight (%)
Fibre		
Cellulose	Crystalline	45
Matrix		
Lignin	Amorphous	20
Hemi-cellulose	Semicrystalline	20
Water	Dissolved in matrix	10
Extractives	Dissolved in matrix	5

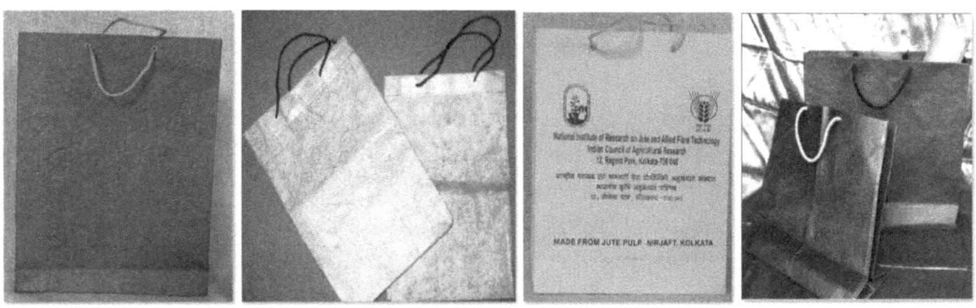

Fig. 2 Different utility bags produced from jute pulp/paper and lightweight jute fabric (see Footnote 14)

5.2 Coated and Laminated Paper/Paperboard for Packaging

Paper is widely used in food packaging because of its advantages of biodegradability and safety for the packaging of food items (El-Wakil et al. 2015). Nevertheless, the porosity and hydrophilic characteristics of paper can easily cause adsorption of water from the surrounding environment or food, thus resulting in loss of the paper's physical and mechanical properties in addition to fostering microbial growth. Hence, the coating/lamination of paper with other materials, such as plastic and metal (aluminium), to address the problems of porosity and hygroscopicity has been quite successful. However, this also decreases the biodegradability and recyclability of the paper. This has ultimately led to the use of biopolymers produced from natural resources as a promising alternative for packaging because they are abundant, renewable, biodegradable, inexpensive, and environmentally friendly. Polysaccharides biopolymers, such as starch, alginates, and chitosan, have been considered for paper-coating (El-Wakil et al. 2015). Lipid compounds, such as long-chain fatty acids and waxes and long-chain alkanes, can also be used as a coating material for paper and paperboard because of their inherent hydrophobic characteristic. Similarly, due to their wide availability, complete biodegradability, good film-forming ability, nontoxicity, good barrier property, and moderate cost, plant proteins have been popularly used for food packaging. It may be mentioned here that compared with whey protein and corn zein, soy protein and wheat gluten have the cost-advantage of being used as affordable packaging materials; however, the mechanical and barrier properties of wheat gluten must be further improved.

Guazzotti et al. (2015) reported the barrier properties of starch-coated paperboard against the migration of n-alkanes and mineral oils. Different types of starches and the presence of sorbitol as plasticizer were tested. For a biopolymeric coating, a 100 % virgin kraft-grade paperboard, unprinted and suitable for dry food, was used. Starch coatings were obtained from maize cationic waxy starch (30 %), maize cationic starch (10 %), or cationic starch mixture with high amylose content based on cereal and tuber starch (10 %). The thickness of the starch coating was in the range of 4.7–14.2 μm. The effective results obtained at 40 °C after 10 days demonstrated the results of the laboratory-scale coatings on paper against n-alkane (range C18–C26) migration compared with uncoated paperboard. Similarly, the use of paper/paperboard as packaging material is still limited because of its inferior water-resistant performance (Chen et al. 2013). Papers and paperboards are the sheet materials comprising an interlaced network of cellulose fibers that is intensely susceptible to water or moisture because it is hydrophilic in nature. To improve functional properties, in many cases an additional barrier coating or lamination of aluminum or plastic (PP or PLA) is incorporated in the paperboard to be used as corrugated box liners (Song et al. 2003; Rhim et al. 2007). However, such laminated packaging material is not only quite expensive, it also has poor recyclability compared with those produced only from paper. This has led to the development of an overprint varnish, another widely used cost-effective and simple protection technology. However, varnish is not a good water repellent. Chen et al. (2013) reported the preparation of a functional overprint varnish to significantly improve the water repellency of paperboard by its unique nanostructured morphology for printed packaging. This is an example of technology derived from biomimicking the water-repellent property of the lotus surface. The functional varnish helps paperboard to utilize its unique lotus-like properties, such as water and moisture-repellency, as well as anti-frost formation, thus resulting in better packaging appearance. The varnish was prepared by mixing silica nanoparticles with an average particle size of 7 nm, decamethylcyclopentasiloxane, polydimethylsiloxane, and nonionic surfactant p-octyl polyethylene glycol phenyl ether at a blend ratio of 3:57:9:1. The water-contact angle reflects an increasing trend of hydrophobicity with the increase of the modifying agent. It was found to reach $\geq$150° in paper packaging when the component of the modifying agent in the varnish was >30 wt%, clearly indicating a super-hydrophobic characteristic (Chen et al. 2013). When the smooth paper of the packaging was coated with unmodified and modified varnish of 20 %, it showed water contact angles of 99° and 110°, respectively. Figure 3 shows the photographs of 10 μL of water droplets on the surface of a packaging paper coated with (a) original varnish and (b) modified varnish comprised of 40 wt% of the modifying agent. Similar to varnish coating, protein can also be used as functional coatings for paper because it acts as a mechanical support for the packaging.

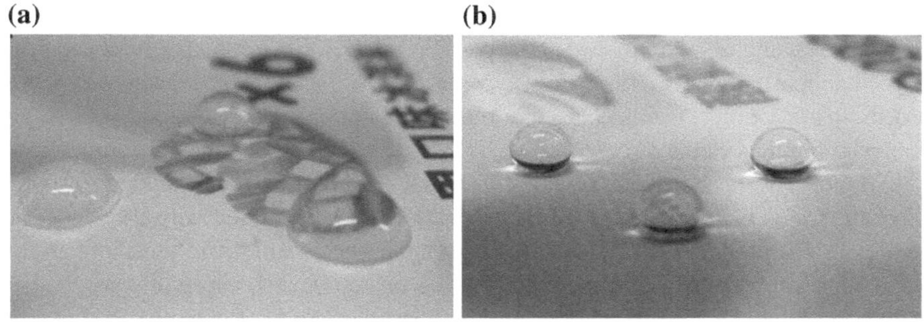

Fig. 3 Photographs of water droplet on the a original and b modified varnish-coated paper (Chen et al. 2013)

In the past, it has been observed that after application of several animal and vegetable proteins—such as caseinates, whey protein isolates (WPI) and concentrates (WPC), corn zein, wheat gluten (WG), and soy protein isolate (SPI) as coating materials of paper and paperboard, they have improved the surface and/or mass-transfer properties of the cellulosic substrate, such as resistance to oil, water, and grease and barrier properties against water vapour, without significantly changing their optical and mechanical properties (Guillaume et al. 2010). Guillaume et al. reported structural and surface features (e.g., water vapour) and gas barrier properties of wheat gluten (WG)-coated paper that could be influenced by certain features of the paper (2010). The transfer properties to water vapour, O_2, and CO_2 of WG-coated papers were significantly improved. Whereas the WG-coating was highly penetrative, the WG-treated paper (WG-TP) behaved as a microperforated material, and the WG-untreated paper (WG-UTP) as a WG film. Water and oil droplets were selected to reflect moisture and grease resistance, which are important criteria for their potential usage as a food packaging material. The water-wettability of WG-coated papers was found to decrease compared with their respective uncoated papers; however, this occurred to a greater extent for WG-TP (approximately 30-fold less) than that for WG-UTP (only 6-fold less).

EI-Wakil et al. (2015) reported the development of a bio-nanocopmosite by casting/evaporation of wheat gluten (WG), cellulose nanocrystals (CNC), and TiO_2 nanoparticles for food-packaging application. A significant improvement in the various properties was observed when 7.5 % CNC and 0.6 % TiO_2 were added to WG. The same composition was chosen to coat a commercial unbleached kraft paper sheet by way of one, two, and three layers of coating. A significant enhancement of 56 and 53 % in breaking length and burst index, respectively, was achieved for paper coated with three layers. This sample exhibited excellent antimicrobial activities, e.g., 100, 100, and 98.5 % against *S. cervisiae*, *E. coli*, and *S. aureus*, respectively, after 2 h of exposure to UVA light illumination. In the increasing domain of the application of sustainable packaging materials, paper and polymer nanocopmosites represent a novel class of packaging materials. Youssef et al. (2013) developed an alternative sustainable and promising material for antibacterial packaging applications. Paper sheets were made from rice straw coated with 5 or 10 % polystyrene (PS) nanocomposites using titanium dioxide nanoparticles (TiO_2-NPs), doped or undoped with silver nanoparticles (Ag-NPs). Polystyrene (PS), a thermoplastic resin with good processing properties, has been used in many applications including food packaging, domestic appliances, electronic goods, toys, household goods, and furniture. Silver (Ag) nanoparticles showed a better antibacterial efficacy than TiO_2 nanoparticles for all of the tested bacteria except for Staphylococcus. Paper coated with PS nanocomposites in the presence of TiO_2 and/or Ag nanoparticles improved tensile strength, water absorption, and air permeability of the coated sheets, especially when they were treated with 5 % TiO_2 without Ag nanoparticles (Youssef et al. 2013). Lavoine et al. (2014) reported a new paper-based packaging with antibacterial efficacy utilizing the synergistic action between beta-cyclodextrin (βCD) and microfibrillated cellulose (MFC). Here, carvacrol (CA), an antibacterial molecule, was incorporated into βCD, which was previously grafted onto the paper substrates by impregnation. The MFC suspension was coated on the ensued substrate surface using a bar-coating process. Due to the presence of citric acid, the grafting process drastically damaged the mechanical properties of the paper substrate, but the air resistance was significantly improved. In the study, synergy between βCD and MFC was established, thus paving the way for the development of a promising technology: sustained-release packaging. Due to grafting of βCD, the paper samples remained antibacterial during the course of 14 h compared with 4 and 6 h for the reference and CA-grafted samples, respectively. Food-packaging materials fundamentally contribute to food quality and safety, as they protect the packaged food against the external sources. In this context, the determination of the hygiene status of the packaging material itself is very important (Feichtinger et al. 2015). However, European legislation neither sets any microbiological criteria, nor provides any approved standard for the microbiological testing of food-packaging materials. Nevertheless, reliable routine control is essential for ensuring the hygiene attribute of the packaging. With the aim to achieve a maximum recovery rate at low-contamination levels, an improved experimental design was developed by Feichtinger et al. (2015). Two different types of paper laminates, a paper—PET laminate (for packaging chilled dairy desserts) and a paper–aluminum laminate (for wrapping chocolate bars) were used as sample materials and were examined for their microbial contamination before and after spiking.

Similarly, starch-, protein-, nano-, and other material-based coatings and, more recently, conducting polymeric coatings, have attracted great attention in the development of new functional papers and packaging materials for their applications as antistatic and electromagnetic-shielding papers, novel wall coverings, electrical-resistant heating papers, and antibacterial papers (Youssef et al. 2012). Most often, polypyrrole (PPy) is preferred as a conducting polymer for fibre coating because of its positive attributes of low toxicity, chemical stability, and easy commercial availability. Further, polyaniline (PANi), another type of well-known conducting polymer, is characterized by its highly p-conjugated polymeric chain, metal-like conductivity, reversible chemical properties, and different morphology as well as electrochemical and physical properties in doping/de-doping process. This has led to the development of many promising end applications. Such coatings can be incorporated into pulp using more than one method, however, the most popular one is the in situ polymerization of a monomer. In addition, the influence of pulp type and the content of acidic groups (e.g., sulfonic or carboxylic groups) was studied. The kappa number (e.g., the residual lignin content) of unbleached kraft pulp and the beating degree of bleached kraft pulp on the conductivity of PANi-coated paper were also studied (Qian et al. 2010). It was found that the amount of PANi coating increased with increasing content of sulfonic groups or decreasing kappa number of the unbleached kraft pulp. Youssef et al. (2012) reported the development and properties of conductive paper based on cellulosic raw materials from unbleached bagasse and/ or rice straw fibers. The hybrid materials can be included in other materials such as plastics, surface coated and anti-static packaging materials or anti-bacterial papers. It was observed that the electrical conductivity (S cm^{-1}) increased profoundly from 8×10^{-13} in the untreated rice straw sample to 2.5×10^{-5} in the 10 % PANi-treated sample. With the increasing application of PANi, the conductivity was found to improve, and a similar trend was also observed in the bagasse sample. However, the breaking length, burst factor, and tear factor were found to decrease with the increasing ratios of PANi added.

Gallstedt and Hedenqvist (2006) investigated different ways of imparting the barrier component at an early stage of the paper-making process to explore the possible synergistic effects of mixing pulp fibers and chitosan in terms of the barrier properties of the sheet and the buffering-induced shrinkage. Paper-making processes were simulated either by using a laboratory paper machine or by solution-casting sheets in Petri dishes. Chitosan–acetic acid salt agglomerated with the pulp fibers and the sheet homogeneity increased during the pressing after sheet formation. The most homogeneous sheets were obtained by solution-casting in the Petri-dish system. At chitosan solution contents >50 wt%, a sufficiently continuous chitosan–acetic acid salt phase was formed, which led to the formation of a sheet with low oxygen permeability. The shrinkage during the buffer treatment could also be effectively reduced due to the presence of pulp fibers. The natural self-assembled micro-structured particles (diatomaceous earth) were used to develop a gas sensor paper with a detection mechanism based on the changes in visible and distinct colour of the sensor paper when exposed to volatile basic nitrogen compounds (Hakovirta et al. 2015). The coating formulation of the paper was prepared by applying diatomite polyvinyl alcohol (PVOH) and pH-sensitive dyes on the acidic paper substrate. The surface coating was designed in such a way as to allow maximum gas flow through the diatomite sensors. The prepared sensor paper, when tested for sensitivity using different ammonia concentrations, exhibited a lower sensitivity limit of 63 ppm.

5.3 Corrugated Boxes in Packaging

Cardboard boxes are industrially prefabricated and primarily used for the packaging of goods and materials. Specialists in the industry seldom use the term "cardboard" because it does not denote a specific material.[15] Corrugated box, a machine-shaped paperboard container with a hollow structure, has been extensively accepted in the packaging and transportation of various goods owing to its advantages of light weight, low cost, ease of assembly and disassembly, good sealing performance, cushioning and antivibration properties, easy recovery behaviour, and amenability to waste treatment (Chen et al. 2011). Their popularity is also due to their good stacking strength in a dry state, easy of availability, and lower cost. Unlike other types of packaging materials, these kinds of boxes protect the packed material from mechanical damage due to drops, impacts, vibration, and compression loads (Pathare and Opara 2014). Corrugated boxes have been extensively used for the transportation and storage of fresh produce in the horticultural industry. Corrugated board has many specific advantages such as low mass (saves money when transporting), easily customizable, strong and stiff, easy to handle, easy to print, and recyclable. Paper and board (38 %) is the mostly used consumer packaging, followed by usage of plastic (30 %), since 1903, when the corrugated box was first accepted by legal freight classification organizations as the containers suitable for freight transportation. China introduced the use of the corrugated box as the external packaging box in the early 1930s. At that time, 80 % of external packing boxes in use were from wooden boxes with cartons accounting for only approximately 20 % of packaging. However, use of corrugated boxes penetrated the market, and by the end of 1940 the percentage of these boxes in use reached to 80 %. Currently approximately 90 % of packing boxes in use are made of corrugated boxes due to technological development in packaging materials and machines. Mainly five different types of corrugated boxes can be produced by various optimal combination of raw materials: (1) lightweight corrugated box, (2) high-strength corrugated honeycomb composite board, (3) intensified sandwich-corrugated cardboard, (4) four-layer corrugated cardboard, and (5) network-structured corrugated cardboard (Chen et al. 2011). A corrugated roll is made of crepe paper, tissue paper, paper boards, plastic films, and many other fibres (see Footnote 15). These products are also widely used for the packaging of products such as plastic ware, glass ware, and steel utensils.

It may be noted that corrugated boxes are predominantly used for export items, whereas reusable plastic containers (RPC) are mainly used in the domestic market. The analysis and prediction of the stacking compression-load capacity of corrugated boxes is important and was studied in detail by Pathare and Opara (2014) to analyze the response of the existing packaging to mechanical stress or to design entirely new boxes to meet postharvest handling conditions. Finite-element analysis and simulation were found to be useful to study and structurally design ventilated corrugated packaging considering the shape, location, and size of the vent. Zhang et al. (2011) described the use of corrugated cartons in packaging low-temperature yogurt with a focus on hazard factors of its external packaging cartons in the logistics process. On the basis of production, storage, and transportation of yoghurt, the factors investigated in their study were relative

[15]http://www.corrugatedboxess.com/corrugated-boxes.html, dated 22-05-2015.

humidity of the storage environment, stock time, storage and stacking, circulation, transportation, handling, and loading and unloading procedures. Similarly, the hazard factors and degree of hazard encountered in each logistics link were also analyzed and summarized.

The most suitable material for packaging fruits and vegetables are wooden and corrugated fibre board (CFB) boxes made of wood pulp and polyethylene films. By considering the requirement of packaging material today, it is not possible to meet such a demand of timber (wood pulp) from the existing forest coverage without causing major ecological imbalance. The use of CFB boxes in India is limited due to the higher cost of forest-based raw materials. However, it is possible to considerably lower the cost of production of CFB boxes by using agricultural and horticultural fibrous or nonfibrous biomass such as cotton plant stalks, jute plant stalk, pineapple leaves, banana pseudo stem, sisal fibres, etc., which are available abundantly from eco-friendly and renewable sources and are not effectively used for other high-valued applications. The current availability of agricultural biomass in India is estimated to be approximately 120–150 million MT including the agricultural and forestry residues. Using these agricultural biomasses to make CFB boxes has an economically viable potential opportunity while preserving the existing forest coverage. Advancements are continually taking place in developed countries pertaining to both the material as well as the art of packing due to the availability of high quality, soft-wood raw material. Among developing countries, China has already been quite successful in the utilization of cotton stalk for manufacturing of CFB boxes. Large amounts of corrugated paper rolls made of crepe paper, tissue paper, paper boards, plastic, films and many other materials are presently used as a biodegradable packaging of plastic ware, glassware, and steel utensils. This kind of specialty single-, double-, triple-layered and composite papers also find application in industries such as food processing, stationery, textiles, plastic, and processed foods, e.g., bakery, confectionery, and breads, due to their positive attributes of high flexibility and smooth finishing. Corrugated rolls are also available in different lengths and colors to fulfill the specific demands of customers.

6 Biodegradable Polymer/Film and Bio/Nanocopmosite for Packaging

6.1 *Biodegradable Polymer or Film*

The current global consumption of plastics is >200 million tonnes with an annual growth of approximately 5 %, thus representing the largest field of application for crude oil. In the last two to three decades, the production and application of plastics, such as PET, PVC, PE, PP, PS, and PA, have increased exponentially worldwide as packaging materials due to their characteristic advantages of availability in large quantity, low cost, good tensile and tear strength, and good barrier properties to oxygen, carbon dioxide, anhydride, and aroma compounds, heat-sealability, thermoplasticity, and inertness (Vigneshwaran et al. 2011). However, so far with the progress of time, waste disposal and management has become a concern, and the situation has worsened because the packaging materials are not biodegradable. As a result, research has intensified on the development of plastics that degrade faster in the environment, thus leading to a complete mineralization or bioassimilation of the plastics (Avella et al. 2005). At present, biopolymers are getting attention regarding those applications where biodegradability and/or the derivation of natural resources is an added value, particularly where valuable petroleum-based plastics are used for applications with a short lifetime. The biodegradable material can be used as a packaging material, such as a shopping bag, which has a shorter lifetime application and where the recycling is either difficult or not economical or both. The term "biodegradable" material is used to describe materials that can be degraded under specific environmental conditions by the enzymatic action of living organisms such as bacteria, yeasts, fungi. The ultimate end products of the degradation process would be CO_2, H_2O, and biomass under aerobic conditions and hydrocarbons, methane and biomass under anaerobic conditions (Avella et al. 2005; Mensitieri et al. 2011). According to the European Bioplastics, biopolymers made from renewable resources must be biodegradable and, especially, compostable so that they can act as fertilizers and soil conditioners at the end of their service life. At present, there is a considerable demand to replace partially or fully synthetic plastic with biodegradable polymers. In this regard, some of the bio-natural polymers such as aliphatic polyester (polycaprolactone) and polylactic acid, have been attempted as packaging materials, but they have not been commercially successful due to their higher cost compared with competitive petrochemical-based polymers. For the sake of simplicity, biodegradable polymers derived from renewable resources are called "biopolymers" in food packaging and for other applications. Currently, biodegradable plastic represents just a tiny market compared with conventional petrochemical-based plastic due to their higher price. Among various biomaterials derived from renewable sources, starch-based products are the most widespread and economically feasible (Avella et al. 2005). To engineer the physicochemical properties of plastics obtained from various renewable resources to improve/meet certain processing requirements and functional and structural demands, several chemicals and additives have been developed to be added to the polymer, e.g., stabilizers, antioxidant, plasticizers, fillers, and processing aids (Mensitieri et al. 2011). Biodegradable polymers can be broadly classified into three groups, namely,

- polymers directly extracted or removed from biomass such as polysaccharides and proteins
- polymers produced by classical chemical synthesis starting from renewable bio-based monomers, such as polylactic acid (PLA), and
- polymers produced by microorganisms or genetically modified bacteria such as polyhydroxyalkanoates, bacterial cellulose, xanthan, and pullulan.

Brief details of some of the best known biodegradable polymers are reported below.

Polylactic acid (PLA): PLA is one of the current versatile biodegradable polymers whose properties, such as the degree of crystallinity, melting, and glass transition temperature, can be tailored by controlling the L and D isomeric forms. The structural, thermal, crystallization, and rheological properties of PLA as well as the specific mechanical processes, such as extrusion, injection molding, injection stretch blow molding, casting, blown film thermoforming, foaming, blending, fibre spinning, and compounding, have been reported in the literature in detail (Mahalik and Nambiar 2010; Lim et al. 2008). PLA shows much lower barrier properties than the well-known polyethylene terephthalate (PET) and it is difficult to heat-seal. However, these challenges could be overcome by blending PLA with other polymers by using micro and nanocopmosites, by coating it with high-barrier materials, and by polymer modification. At present, it is possible to produce plastic shopping bags from polylactic acid (PLA), a biodegradable polymer derived from lactic acid, a vegetable-based bioplastic (Shah et al. 2008). This material biodegrades faster under composting conditions and does not leave a toxic residue.

Cellulose: Cellulose is extracted chemically by isolation from its crystalline state in microfibrils. It is fusible and soluble in hydrogen bond-breaking solvents such as *N*-methylmorpholine-*N*-oxide. Under normal conditions, because of its infusibility and insolubility to others, its derivatives have been explored for packaging applications (Mahalik and Nambiar 2010).

Protein based film: Several protein sources have been proposed for the preparation of new thermoplastics. Protein-based films can also act as barriers to oxygen, carbon dioxide, and oil and fats, which are the important desirable properties of a packaging material. On the other hand, their mechanical and water-vapor barrier properties are generally inferior to those of synthetic-originating materials (Mensitieri et al. 2011). Among the various proteins suitable for film formation only zein, a prolamine from corn, has been extensively studied for research and industrial applications owing to its unique hydrophobic characteristic due to the presence of high content nonpolar amino acids. Zein possesses a substantially better moisture-barrier property than any other proteins such as casein or polysaccharides such as starch.

Poly-beta-hydroxyalkanoates (PHB): PHB, a member of poly hydroxyl alkanoates, degrades in the presence of various microorganisms that in contact of polymer secrete enzymes, which is responsible for breaking the polymer into smaller parts (Sorrentino et al. 2007). The PHB is 100 % resistant to water, biodegradable, and thermoplastic in nature. Typically, it is a highly crystalline thermoplastic with very low water-vapor permeability akin to the low-density polyethylene (LDPE). However, it suffers from a major drawback of unfavourable ageing process during application.

Polyhydroxy-co-3-butyrate-co-3-valerate (PHBV): Among the matrices used for the preparation of biocomposites. PHBV, a bacterial aliphatic copolyester, has been reported to be produced from by-products of the food industry, and it possess the complete biodegradability during composting, backyard, or landfill conditions, and/or recyclability (Berthet et al. 2015). It could be easily processed through extrusion or injection. It displays a high water-barrier property and has acceptable mechanical properties despite of tendency toward brittleness. However, it is still costlier for food-packaging applications; in addition, its barrier properties are not sufficient enough for fresh foods such as cheese, fruits, or vegetables that require respiration packaging.

Starch polymer: Starch is a semi crystalline polymer stored in granules in most of the plants. It is composed of repeating 1,4-α-D glucopyranosyl units of amylose and amylopectin. Whereas the amylose is almost linear in which the repeating units are linked by α–(1–4) linkages, the amylopectin has α–(1–4)-linked backbone with 5 % of α–(1–6)-linked branches. The relative amounts of amylose and amylopectin depend on the plant source. The ratio of the two components characterises the different properties of the material. For example, corn starch granules typically contain approximately 70 % amylopectin and 30 % amylase (Mahalik and Nambiar 2010). In the food-packaging application, starch-based material has received considerable attention due to its advantages of biodegradability, low cost, renewable in nature, thermoplasticity, and wide availability at a much lower cost (<1 euro/kg) (Vigneshwaran et al. 2011; Avella et al. 2005; Mensitieri et al. 2011). Biodegradation of starch-based polymers occurred due to enzymatic attack at the glycosidic linkages between the sugar groups, which leads to a reduction in chain length and splitting out into lower molecular-weight sugar units (Mahalik and Nambiar 2010). This holds great promise for application in packaging as an alternative to synthetic nonbiodegradable polymers. It may be noted that starch, as a packaging material alone, does not form films with adequate mechanical properties until and unless it is first plasticized or chemically modified. Additionally, it is still not being used to its fullest potential due to the limitation of possessing strong hydrophilicity, poor moisture barrier, high brittleness, and inadequate mechanical properties (Azeredo 2009).

6.2 Reinforcement of Biodegradable Polymer

As described earlier, the use of synthetic petrochemical based polymers, such as PET, PVC, PE, PP, PS and PA, has been partially restricted because they are not fully recyclable and/or biodegradable and hence pose a serious threat to ecology (Vigneshwaran et al. 2011). Awareness of the waste-disposal problem and its impact on the environment has created a new interest in the area of degradable polymers (Shah et al. 2008). Several natural polymers—such as cellulose, starch, lignin, and chitosan (carbon-based polymers)—are biodegradable and compostable. Despite several technological interventions in biopolymers for eco-friendly food packaging, their utilization has still not reached their full potential due to poor mechanical and barrier properties and greater cost. These limitations have recently been partially addressed by incorporating micron- to nano-sized reinforcing ingredient (fillers) during composite preparation. The poor barrier to gases and vapors and poor mechanical properties of biopolymers have led to newer research and development (R&D) in improving these properties. R&D in polymeric materials, appropriate fillers, matrix and filler interaction, and new formulation strategies to develop composites have potential opportunities in food-packaging application (Majeed et al. 2013). Polymer composites are mixtures of polymers with inorganic or organic additives having certain geometries such as fibres, flakes, spheres and particulates. Natural fibres as bio-fillers have been the preferred choice because they exhibit advantages such as low cost, low density, reduced tool wear, and acceptable specific strength in addition to their renewable and degradable characteristics. In most cases, bio-fibres are cheaper than synthetic fibres and cause fewer health and environmental hazards compared with glass fibre-based composites. This may lead to the production of highly durable consumer products from natural fibres that can be easily recycled. Polyhydroxy-co-3-butyrate-co-3-valerate (PHBV) is a completely biodegradable matrix used for the preparation of biocomposites. However, it has an inadequate barrier property, an important requirement in respiring packaging, but it was improved by the addition of other fibres. Berthet et al. reported the impact of wheat-straw fibre size, morphology, and content on the mechanical properties and water-vapour permeability of PHBV-based composite materials (Berthet et al. 2015). Three types of wheat-straw fibres varying in diameter of 17, 109, and 469 μm were used. Based on the various analyses, it was postulated that the new range of PHBV-based composites with tunable properties could be developed successfully as per the requirements of respiring fresh-food products such as strawberries, thus enabling to preserve them in a better way compared with the current use of polyolefins.

In the process of manufacturing biodegradable packaging materials, many functional as well as reinforcing additives are used to improve their physical, mechanical, barrier, thermal, and functional properties. In this regard, it has been observed that incorporation of PLA isomers, sodium caseinate, and whey protein decreases the glass-transition temperature of the polymer and increases tensile strength and puncture resistance. Similarly, zein, a major component of corn protein, has become an important industrial material providing biodegradablity as well as good tensile and water-barrier properties. In addition, zein in nano-form, such as nanobeads or nanoparticles, can be used as edible carriers for flavor compounds or for the encapsulation of nutraceuticals as well as to improve the strength of plastic and bioactive food packages.

6.3 Nanocompsite and Biocomposite for Packaging

In the last decade, due to the rapid growth of nano-science and nanotechnology, it has been proven that the application of organic/inorganic nanomaterial as filler could enhance the mechanical and barrier properties of various nanocopmosites. The applications of nanotechnology in the agriculture and food sectors are relatively recent compared with their use in drug delivery and pharmaceuticals (Sozer and Kokini 2009). The application of nanotechnology in polymers may open new possibilities for improving not only their properties but also their cost-efficiency

(Azeredo 2009). In this context, it is worth mentioning that the application of organic and inorganic nanomaterials (e.g., clay, silica, cellulose, chitosan, starch nanocrystal, zein-nano beads, TiO_2, and silver) in film or matrix (e.g., starch, carrageenan, cotton seed protein, and soya protein) has shown improvement in mechanical, thermal, gas-barrier, antimicrobial activity, enzyme immobilization, biosensing, oxygen scavenging properties, etc. (Azeredo 2009; Sozer and Kokini 2009). The common plasticizers for hydrophilic polymers are glycerol, polyether, urea, and water. Starch as a film or bag could be employed in packaging fruits and vegetables, snacks, or dry products (Savadekar and Mhaske 2012). Efficient mechanical, oxygen, and moisture protection is desirable in such applications. Thermoplastic starch (TPS) alone often cannot satisfy all these requirements because its hydrophilic nature that likely causes change in thermoplastics' performance during and after processing due to changes in the water content. In addition to the improvement in mechanical and barrier properties of the packaging materials due to the incorporation of nano-fillers, various nano-structures are also responsible for providing active or "smart" properties, such as antimicrobial activity, enzyme immobilization, biosensing, etc., to the packaging system (Azeredo 2009). The use of nano-scale fillers leads to the development of polymer nanocomposites with improved tensile modulus, dimensional stability, and resistance to solvent or gas. Such composites also possess additional benefits such as low density, transparency, good flow, better surface properties, and recyclability by the addition of a filler (Sorrentino et al. 2007). The massive effort to extend shelf life and enhance food quality while reducing packaging waste has encouraged exploration of new bio-based packaging materials such as edible and biodegradable films from renewable resources (Sorrentino et al. 2007). Also, the quality of packaged food is directly related to the attributes of food as well as the packaging material. Most of the food's qualities become deteriorated due to mass-transfer phenomena such as moisture absorption, oxygen invasion, flavor loss, undesirable odour absorption, and migration of packaging components in the food (Galic et al. 2011). The phenomena can occur between the food product and the surrounding atmosphere, i.e., between the food and the packaging materials, or among the heterogeneous ingredients in the food product itself. Thus, the rate of transport of such reactants across the partial barrier of the package wall can become the limiting factor in the shelf life of the packed food. To increase the shelf life of processed foods, the package must be designed in such a way as to have adequate water-vapor (WV) and/or gas (O_2, CO_2, etc.) permeability. The use of nanocomposites in food packaging is attracting considerable interest due to their many fascinating features as discussed below (Majeed et al. 2013).

(i) ***Nano cellulose and its starch based composite***

Though the use of plastics in food packaging and agriculture is essential, plastics have simultaneously adverse medium to long-term effect in polluting soil, food and the environment. On the other hand, indeed many of the available alternative biopolymers, such as starch and k-carrageenan, cannot compete techno-mechanically with the well-established synthetic counterpart due to their inadequate mechanical and barrier properties (National Agricultural Innovation Project 2012). Presently, nonspinnable short cotton fibres and cotton linters, being a major source of cellulose and ubiquitously available in India, has found application in the production of microcrystalline cellulose, cellulose powder, and cellulose acetate to be used as filler in nanocopmosites and other high-value, low-volume applications. Cellulose is one of the most important, abundant, renewable, and biodegradable natural polymers and it exists in the biomasses of several plants, such as wood, cotton, hemp, straws, sugarcane bagasse, and other plant-based materials. It has a wide range of applications in the form of fibre, paper, films, and polymers. The utilization of such natural biomass in novel applications has recently attracted the global interest due to its ecological and renewable characteristics (Li et al. 2012). Basically two types of nano-reinforcements, such as microfibrils and whiskers, can be obtained from cellulose. In plants or animals, cellulose chains are synthesized to form microfibrils (or nanofibres), which are bundles of molecules held together through hydrogen bonding (Azeredo 2009). Depending on their origin, microfibrils have nano-sized diameters of 2–20 nm and lengths in the micrometer range. Each microfibril is formed by an aggregation of elementary fibrils made of crystalline and amorphous parts. The crystalline parts isolated by different treatments are called "whiskers," also known as "nanocrystals," "nanorods," or "rod-like cellulose microcrystals," and have a high aspect ratio with a diameter of $\leq$8 to 20 nm and lengths ranging from 500 nm to 1–2 μm. In general, nanocrystals of cellulose with diameters ranging from 2 to 20 nm and length ranging from 100 to 2.1 μm (more precisely <100 nm for defect free-crystal) have different names in the literature such as "cellulose nanowhiskers," "cellulose whiskers," whiskers, "nanowhiskers," "nanofibrils," "nanofibres," "cellulose crystallites," "cellulose crystals," "cellulose nanocrystals," "nanocrystalline cellulose," "cellulose monocrystals," and "cellulose microcrystals." These are possible to produce from cotton linters as well as many other natural fibrous agro-biomass (Morais et al. 2013; Cherian et al. 2011; Neto et al. 2013). Over the years, due to the advancement in nano-science and technology along with nano-scale characterization techniques, attempts have also been made to produce microcrystalline cellulose, nanocellulose particles, and nanocellulose fibres to be used as a reinforcing agent in different nano-/bio-composites due to their unique advantages of superior physical properties and environmental benefits such as large specific surface area (estimated to be several hundreds of $m^2\ g^{-1}$), very high modulus of elasticity ($\approx$150 GPa), and high aspect ratio, thus ensuring high strength with low-filler loading, low density ($\approx$1.56 g/cm^3), nonabrasive and nontoxic nature, biocompatibility, biodegradability, and being produced from renewable agro biomasses at a lower cost (Azeredo 2009; Neto et al. 2013). They can also be used as reinforcements for adhesives, components of electronic devices, biomaterials, foams, aerogels, and textiles (Morais et al. 2013). In the literature, a number of approaches have been reported for the production of highly purified nanocellulose from various cellulosic to ligno-cellulosic biomasses such as cotton linter, cotton fibre, sugarcane bagasses, pineapple leaf, soy hulls, and corncob (Li et al. 2012; Morais et al. 2013; Cherian et al. 2011; Neto et al. 2013; Silverio et al. 2013; Santos et al. 2013; Rosa et al. 2010; Haafiz et al. 2014). The methods included are steam-explosion treatment, acid or alkaline hydrolysis, enzyme-assisted hydrolysis, microbial process, high-pressure homogenization, as well as a combination of two or several of the aforementioned methods.

Vigneshwaran et al. reported the production of nano-cellulose by microbial, enzyme and mechanical process consisting of refinement and homogenization, and a comparative assessment was also made between such processes (National Agricultural Innovation Project 2012). The same research group reported the application of nanocellulose as a reinforcing agent in starch film. It was observed that due to its high surface energy as well as high hydrophilic characteristic, nanocellulose tends to aggregate during film formation. This problem was addressed by the addition of gum arabic to assist in the uniform distribution of nanocellulose. The nanocellulose–starch film was prepared by using a soluble starch derived from potatoes by acid hydrolysis to a consistent molecular weight. The film-forming solution was prepared by gelatinizing the starch (4 %) at 95 °C followed by the addition of 0.02 % sodium azide and 0.5 % glycerol antimicrobial and plasticizing agent. In the composite-film preparation, the nanocellulose add-on was kept at 1 % of the weight of starch. In an alternative method, a solution-cast film of k-carrageenan was prepared using 0.1, 0.2, 0.3, 0.4, 0.5, and 1 % nanocellulose fibres using distilled water as the solvent. A similar sample of k-carageenan was also prepared in the presence of nano-cellulose instead of nanocellulose fibres. The physical and mechanical properties of starch-nanocellulose composite are reported in Table 5.

Similarly, Savadekar and Mhaske (2012) reported the improvement in mechanical properties of starch films by the incorporation of nanocellulose fibres (NCF). The NCF were successfully synthesized from short staple cotton fibres by a chemo-mechanical process, and its composite with thermoplastic starch (TPS)

Table 5 Different properties of biodegradable starch films (Vigneshwaran et al. 2011; Avella et al. 2005; National Agricultural Innovation Project 2012)

Different film	Tensile strength (MPa)	Elongation at break (%)	Thickness (μm)	Surface energy (dyne/cm)	WVTR (g/m^2h^1)	Solubility (%)
Control starch film	1.35 ± 0.8	20.2 ± 2.5	140 ± 2	40.5 ± 2.5	388 ± 15	39.5 ± 0.5
Starch + nanocellulose	3.27 ± 1.1	22.8 ± 2.9	150 ± 2	28.7 ± 2.3	265 ± 13	35.7 ± 1.7
Starch + nanocellulose + gum arabic	4.79 ± 1.3	36.6 ± 3.8	150 ± 2	21.2 ± 2.2	181 ± 10	32.1 ± 0.5
	Stress at peak (MPa)	Elongation at break (%)	Young modulus (MPa)			
Starch	19	3	979			
Starch + clay (4 %)	22	4	1135			

WVTR Water vapor transmission rate

was prepared by solvent-casting method. The 0.4 wt% NCF-loaded TPS films showed 46 % improved tensile strength compared with the base polymer film, but beyond 0.5 wt% the addition of NCF was found not to be beneficial because the tensile strength started to deteriorate. Oxygen permeability was found to decrease significantly, i.e., by 93 % in the 0.4 % NCF/TPS sample compared with the control TPS sample, possibly due to increased tortuous pathways used for the permeation of oxygen molecules in the presence of NCF in starch film. Similar result were also observed in the water-vapor permeation rate. In a similar experiment, the moisture barrier of polymer films was observed to improve owing to an increase in the tortuosity in the materials, thus leading to slower diffusion processes and hence lower permeability (Azeredo 2009). The barrier properties are expected to enhance if the filler with a high aspect ratio is uniformly dispersed in the matrix. It was seen that the incorporation of nanocellulose as a filler could increase the tensile strength of starch film by 3.5 times. A similar result was also observed in a water-vapour permeability test where permeability was found to decrease by 2 times (Vigneshwaran et al. 2011). It was also observed that oxygen permeability was decreased by 93 % in the case of a nanocomposite film compared with a control starch film. Figure 4 shows the packaging of strawberries and broccoli in a starch/nanocopmosite film. In the biodegradability study, the starch–nanocellulose composite film was found to degrade in 21 days by the native microbial population in garden soil. From the above-mentioned observations, it can be inferred that this kind of composite film may be suitable for food packaging and agricultural field-mulching applications.

Similar to the application of starch/nanocellulose fibre/particle or starch/clay composite film, Savadekar et al. (2012) reported a structure-property evaluation of kappa-carrageenan (KCRG) and nano-fibrillated cellulose (NFC) composite film for application in food packaging. Carageenan is a water-soluble linear

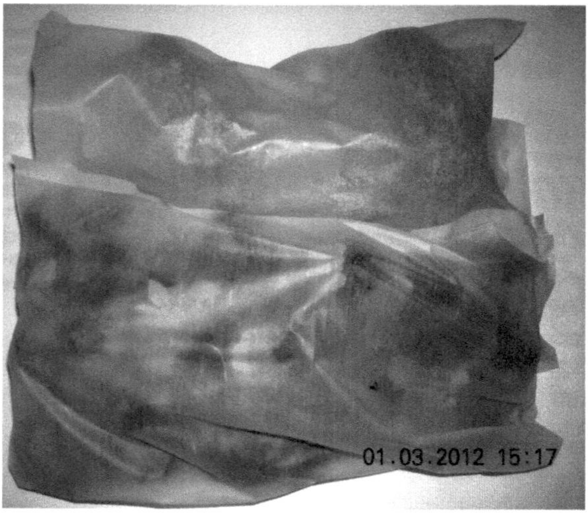

Fig. 4 Packaging of strawberries and broccoli in starch–nanocellulose composite film (National Agricultural Innovation Project 2012)

polymer (polysaccharides) extracted from red seaweed and extensively used in foods, cosmetics, and pharmaceuticals. This particular film has the advantages of good transparency, tensile strength, gelling ability, and film-forming capability. However, it suffers from higher cost, poor barrier properties, and lower tensile-breaking elongation. Carageenan as edible films and coatings has already been used in the food industry for the packaging of fresh and frozen meat, poultry, and fish to prevent superficial dehydration and for other purposes. The NFC was prepared from short staple cotton fibres by chemo-mechanical process in a laboratory disc refiner. The diameter of fibril under SEM was estimated to be 242 ± 158 nm. It was seen that the tensile strength of KCRG increased with increasing NFC loading up to 0.4–0.5 wt% followed by a decrease. Similarly, the 0.4 % nano-fibrillated cellulose showed the lowest water-vapor and oxygen-transmission rates, i.e., approximately 80 % reduction compared with the control film (Savadekara et al. 2012). In 1993, LDPE-starch blends were commercialized under the trade name Ecostar (Siracusa et al. 2008). Other commercial trade names of LDPE-starch blends are Bioplast® (from Biotec GmbH) and NOVON® (from NOVON International) (Siracusa et al. 2008).[16] Salehudin et al. reported the development of a starch–chitosan hybrid film that is totally degradable because it is produced from a renewable material. Its low mechanical properties were improved by the addition of oil palm empty fruit bunch (EFB) cellulose nano-fibres (Salehudin et al. 2014). The role of chitosan in the starch film packaging was to ensure the killing of pathogens (antimicrobial) and hence to increase food shelf life. Transmission electron microscopy (TEM) images showed nanofibre diameters in the range of 1 to 100 nm. Nanocopmosite film was constructed by keeping the cellulose nanofibre content, constant at 2, 4, 6, 8, and 10 % weight of the starch. The tensile strength of the control starch–chitosan film was 3.96 MPa, and it was increased to the highest value of 5.25 MPa in the 8 % nanocellulose-incorporated sample. The antimicrobial efficacy result showed that the addition of cellulose nanofibre could increase the inhibition effect toward gram-positive bacteria but not toward gram-negative bacteria.

Dogan and McHugh (2007) reported that microcrystalline cellulose (MCC) with submicron-size diameters had a much higher effect on tensile strength in hydroxyl propyl methyl cellulose (HPMC) than their micron-sized MCC counterpart Additionally, the negative impact of the micron-sized MCC on the elongation of the films was much more noticeable than that of its submicron-sized counterpart. Nanocomposites of pea starch matrix with cellulose whiskers extracted from pea hull fibers have also reported. The composite showed the highest transparency and the best tensile properties when they were produced from whiskers with the highest aspect ratio (Chen et al. 2009). The thermal stability of the polymers in nanocomposites with cellulose whiskers was reported to be greater than those made from corresponding bulk polymers (Azeredo 2009).

[16]www.designinsite.dk, dated 22-05-2015.

(ii) *Starch and various nanofiller-based composite*

The main challenge of preparing nanocopmosites is the uniform dispersion of nanofiller and in the present case, nano-clay in the biopolymer matrix. Montmorillonite is the most commonly used natural clay that has been successfully applied in numerous nanocomposite systems (Savadekar and Mhaske 2012). Avella et al. (2005) reported the novel biodegradable starch/clay nanocomposite film preparation and their requisite property evaluation to be used as food packaging. The films were made by the casting process using potato-based starch and 4 % purified montmorillonite clay. In some of the films, biodegradable polyester was also added. All of the mechanical properties were evaluated by storing the samples in three different relative humidity conditions to correlate the effect of different moisture conditions during storage and the influence of water presence on the final performance of the films. It was clearly observed that the presence of clay profoundly increases the young modulus of the starch film at all the humidity conditions under the experiment. On the other hand, the film samples to which biodegradable polyester was added showed a decreases in young modulus irrespective of the humidity conditions. At low humidity, the samples showed better mechanical properties. It seems that water strongly affects the modulus of the starch blends. In the same study, the researchers also reported the presence of metal particles in the film, which can migrate and come into contact with the food. Therefore, analysis of some vegetables (lettuce and spinach) in contact with the starch-based biodegradable films was performed and no significant increase of iron (Fe) and magnesium (Mg) was found in the vegetables. However, a slightly higher silicon (Si) content was observed, which was possibly due to the presence of clay nanoparticle containing silicone (Avella et al. 2005). Based on the actual regulations and European directives on biodegradable material assessment, starch–clay nanocomposite films can be effectively utilized in the food-packaging sector owing to their low overall migration limit.

The addition of <5 % clay as a filler was done to improving the tensile and elongation properties of thermoplastic starch (Sorrentino et al. 2007). Beside, the decomposition temperature was increased, whereas the relative water-vapor diffusion coefficient of TPS was decreased. Similarly, silica nanoparticles ($nSiO_2$) have also been used to improve the chemical and/or barrier properties of several polymer matrices. In this regard, the improvement in tensile properties was reported for a starch matrix due to incorporation of $nSiO_2$ (Azeredo 2009). To increase the barrier properties of zein polymers, it was modified with stable silicate complexes of montmorillonite, hectorite and saponite of 1 nm in thickness, with diameters ranging from 30 to 2000 nm to improve their strength and stiffness, and water and gas permeability, even at low levels of 1–5 vol.% application (Yoshino et al. 2002).

(iii) *Other nanocopmosites*

It is interesting to note that not only starch can be used as a matrix in biodegradable film formation, it can also be used in nanoparticulate form to improve various physico-mechanical properties of different composite materials (Azeredo 2009; Mensitieri et al. 2011). Native starch granules can be submitted to an extended-time hydrolysis at temperatures below the gelatinization temperature, thus enabling the hydrolysis of amorphous regions and resulting in the separation of crystalline lamellae, which are more resistant to hydrolysis. Starch crystalline particles show platelet morphology with thicknesses of 6–8 nm. It has been reported that the tensile strength and modulus of pullulan film can be enhanced by the addition of starch nanocrystal (Azeredo 2009). The water-vapor permeability of pullulan films was decreased by the addition of $\geq$20 % of starch nanocrystal. In a similar vein, the preparation of chitin/chitosan nanoparticles, 500 nm in length and 50 nm in diameter, has been obtained by acid hydrolysis of chitin. Lu et al. incorporated chitin whiskers into soy protein isolate (SPI) thermoplastics, and it was observed that the whiskers greatly improved the tensile strength, elastic modulus, and water-resistance properties (Lu et al. 2004). Zein, a prolamin and the major component of corn protein, has recently been an important material for food-packaging applications owing to its unique properties. Biodegradable zein films with good tensile and water-barrier properties were prepared by dissolving zein in either ethanol or acetone (Sozer and Kokini 2009; Yoshino et al. 2002). Zein nanobeads or nanoparticles can also be used as edible carriers for flavor compounds or for the encapsulation of nutraceuticals as well as to improve the strength of plastic and bioactive food packages (Sozer and Kokini 2009). Similarly, silicates consisting of crystalline layers with 1-nm thickness and diameters ranging from 30 to 2000 nm in nanocomposites are able to control the gas-diffusion rate through their tortuous pathway.

7 Environmental Implications of Biodegradable Packaging

Because packaging waste constitutes a significant portion of municipal solid waste, it has in recent years heightened environmental concerns, thus leading to the strengthening of EU regulations to reduce the quantum of packaging waste (Davis and Song 2006). A large numbers of oil-based polymers, such as PET, PP, PE, PS, PA, and others, are currently being used in packaging applications. These synthetic polymers are nonbiodegradable and also difficult to recycle or reuse due to their admixture with contamination, complex composites, and the presence of different processing additives such as fillers, dyes/pigments, and plasticisers as well as coating or multilayer composite structure used to enhance the product's aesthetic and functional performance. These altogether pose difficulties in collecting, identifying, sorting, transporting, cleaning, and reprocessing of plastic packaging materials, thus making recycling uneconomical; however, disposal to a landfill is a more convenient alternative. With the recent development in biodegradable packaging materials from renewable natural resources with properties more or less similar to those of synthetic polymers, it is anticipated that biodegradable polymers would contribute toward the development of sustainable packaging materials. Davis and Song (2006) reported the impact of biodegradable packaging materials on waste management in terms of landfill, incineration, recycle/reuse, and composting with respect to oil-based polymer packaging materials. In the study, it was observed that biodegradable packaging materials are most suitable for single-use disposable applications, whereas postconsumer used packages can be locally composted as a means of recycling the materials. Establishment of an appropriate collection, transportation, and treatment technologies are considered crucial for the success of widespread applications of biodegradable packaging materials. Because recycling is energy-expensive, compostability is one of the most important attributes of a biopolymer that allows disposal of the packages in the soil. Approximately 67 million tonnes of packaging waste are generated annually in the EU comprising approximately one third of all municipal solid waste (Davis and Song 2006). In the United Kingdom, 3.2 million tonnes of household waste produced annually comes from packages, which equates to >12 % of total household waste produced. In developed countries, food packaging represents 60 % of all packaging. Low-value soiled packaging films and carrier bags may be used as an economic boiler fuel (i.e., incineration combined with electricity generation) because they are usually manufactured from polyethylene and has a very high calorific value. Biopolymers, such as natural fibres and starches, have relatively lower gross calorific value (GCVs) compared with synthetic polymers. However, GCVs values of the biopolymer are close to those of wood, and thus they are still suitable for incineration. Composting is an essential process to breakdown waste through biodegradation, and this is considered the most attractive route for the treatment of biodegradable packaging waste. Usually biodegradable polymers degrade by the same mechanisms as organic matter within aerobic composting systems. The trigger for degradation could be a microbial, hydrolytically, or oxidative susceptible linkage built into the backbone of the polymer, or, alternatively, additives that catalyze the breakdown of polymer chains. This trigger may be specifically designed so as to ensure that degradation does not occur within the "use lifetime," but it should ensure degradation on disposal within a given environment. Although the incorporation of nanomaterials into polymer/film packaging materials is important, it must be noted that nanomaterials, due to their increased surface area, might have adverse effects on humans and animals (Sozer and Kokini 2009). There might also be potential and unforeseen risks associated with their use in food-packaging materials. Because no regulation that specifically control or limit the production and/or application of nanosized particles presently exists, they must be used very carefully.

Carbon footprint (CF) is a measure of the impact of human activities on earth and the environment. More specifically, it relates to climate change and the total amount of greenhouse gases produced as measured by carbon dioxide emission. Muthu et al. in 2011 reported an exhaustive carbon footprint (CF) analysis of recycled and reused plastic, paper, and nonwoven and woven bags sent to land-filling in countries such as China, Hong Kong, and India (Muthu et al. 2011). The first stage of the study, e.g., the baseline study, showed the impact of different types of shopping bags in the manufacturing phase without considering their usage and disposal. In the next stage, the study of the carbon footprint of these bags, including their usage and disposal phases (i.e., cradle-to-grave stage), was measured. The results showed that high CFs of different types of shopping bags if no usage and disposal options were provided. However, the CF values were lower in the case where a higher percentage of reuse was preferred over recycling and disposing in landfill. It is interesting to note that reuse could significantly scale down the carbon footprint. Once the shopping bags have reached the end of their service life, they must be recycled rather than disposed of in landfill. It has been observed that in India, the greenhouse gas emission was the least (708 g) for the polypropylene nonwoven bag and the greatest (3410 g) for the paper bag. For the functional unit assumed, the nonwoven bags consumed less energy and fewer quantities of materials, and less greenhouse gas is emitted in the production phase of shopping bags compared with its counterparts in China, Hong Kong, and India. Reusable bags, such as nonwoven bags made of polypropylene followed by woven cotton bags, seem to be environmentally friendly compared with conventional plastic and paper bags for the functional unit assumed in the comparative study. In this context, and regarding consumer behavior and governmental policies, it is important to encourage people to choose reusable bags and to promote more recycling systems to scale down the environmental impacts made by any type of shopping bags. In another study, the plastic bag was found to be a little better in terms of environmental impacts compared with paper bags (Muthu et al. 2009).

8 Present Status of Biodegradable Packing

The Indian Packtech segment is expected to grow at a rate of 22 % to USD $11,782 million by 2016 to 2017 as per estimates of the Working Group on Textiles and Jute Industry, Ministry of Textiles, Government of India (see Footnote 9). In India, as per government norms, a minimum of 90 % of food grains and 20 % of sugar total production are to be packed in jute fibre-based hessian and sacking bags. Natural fibre such as jute is suitable for the packaging of sugar and other agricultural food grains due to its advantage of low cost, biodegradability, and being produced from renewable sources. The role of the jute industry in Indian economy is important because India is the largest producer of jute globally and the second-largest exporter of jute goods, which ultimately supports the 40 lakh farm families' livelihood. To develop alternative materials to petrochemical-based nonbiodegradable plastics, in recent years the development of biodegradable packaging materials from renewable natural resources has received widespread government support in EU countries, thus resulting in the establishment of many national or international research organizations to facilitate research and development in this area. Some of these include the European Renewable Resource Materials Association, National Non-Food Crops Centre in the United Kingdom, International Biodegradable Polymers Association and Work Groups in Germany, and Interactive European Network for Industrial Crops Application (Davis and Song 2006). The United Kingdom Government-Industry Forum has also strongly recommended greater use of nonfood crops for biodegradable-packaging applications. The development of biodegradable packaging would ensure utilization of crop over crude oil and integrated waste management so as to reduce landfill. To improve the packaging standards in India, the Indian Institute of Packaging (IIP), a national apex body, was set up in 1966 by the packaging and allied industries and the Ministry of Commerce, Government of India.[17] The other objectives of the institute are to promote an export market by way of innovative package design and development, as well as to upgrade the overall standard of packaging in the country. In addition, the ITC's packaging and printing business is the largest converter of paperboard packaging in South Asia. It converts >70,000 tonnes of paper, paperboard, and laminates per annum into a variety of value-added packaging for foods and beverages, personal products, cigarettes, liquor, and consumer goods.[18] The division that was set up in 1925 as a strategic backward integration for ITC's cigarette business, which is now India's most sophisticated packaging house. State-of-the-art technology, world-class quality, and a highly skilled and dedicated team have contributed to the position of ITC as the first-choice supplier of high value-added packaging materials. Recently microcrystalline cellulose and nanocellulose particles/fibres have become the promising fillers in nano/bio-composite due to their many unique properties as discussed earlier. Vigneshwaran et al. reported the production of nanocellulose by microbial, enzyme, and mechanical processes consisting of refinement and homogenization (Annual Institute Report of Central Institute for Research on Cotton Technology 2013; see Footnote 19). Nanocellulose has been used as a reinforcing agent in a starch-based composite film for packaging and other applications. Now, keeping in mind the future commercial application, the Central Institute for Research on Cotton Technology, Mumbai, India, under the Indian Council of Agricultural Research (ICAR), has already taken a lead role in setting up the nation's first pilot plant for the production nanocrystalline cellulose and nano-fibrillated cellulose from cotton linters and other agro-products by chemo-mechanical and microbial processes with a production capacity of 10 kg/d. The products are intended to be used as filler in polymer composites and other technical applications (Annual Institute Report of Central Institute for Research on Cotton Technology 2013; see Footnote 19).[19]

9 Summary

Packaging materials play an important role in providing protection to goods from physical damage, contamination, and deterioration as well as providing information about the product. It also provides sales appeal, consumer convenience and safety, physical barrier from the external environment such as light, oxygen and humidity, a certain degree of cushioning, and antivibration effects. The use of paper pulp in packaging has become more attractive than the traditional materials,

[17]http://www.iip-in.com, dated 22-05-2015.

[18]http://www.itcportal.com/businesses/packaging.aspx, dated 22-05-2015.

[19]www.nanocellulose.in, dated on 03-05-2015.

such as expanded polystyrene foam, due to its advantages of low price of the recycled paper, low cost of production, and biodegradability. Such materials have primarily been used for various flexible, semi-rigid, or rigid packaging of electronic items, food, and other products. Similarly, corrugated boxes have been used since the early 1930s as external packing in freight transportation; at present, approximately 90 % of packing boxes in use are made of corrugated boxes. As per government norms, a minimum of 90 % of the food grains and 20 % of the sugar produced in India are to be packed in jute fibre-based hessian and sacking bags. A natural fibre such as jute is most suitable for the packaging of sugar and other agricultural food grains owing to its advantage of low cost, biodegradablity, and eco-friendliness. In addition, jute is also produced from renewable sources and could satisfy the standard for safe packaging. Textile-based packaging, at on end, includes heavyweight, densely woven fabrics (e.g., bags, sacks for storage); on the other end, it includes lightweight nonwoven fabrics used for durable papers, tea bags, shopping bags, and industrial product wrappings. During the past 50 years, synthetic polymers, along with the advancement in polymer science, material science, and packaging technology, have been steadily replacing traditional packaging materials such as paper, glass, metals, fabric, etc. owing to their advantages of low cost, low density, chemical inertness, flexible to rigid structure, and transparency. However, because they are not fully recyclable and/or biodegradable, they pose serious threats to the environment. This fact has led to the development of biodegradable polymers/films such as starch, polylactic acid, protein-based film, and poly-beta-hydroxyalkanoates. Because the majority of such polymers do not possess adequate physical, mechanical, barrier to gas and vapour, and other functional properties, they were engineered by incorporation of micron- to nano-sized filler or reinforcing agents such as silver, titanium dioxide, chitosan, cellulose, clay, starch, silica, and zein protein. In addition, the development of biodegradable packaging materials from renewable natural resources has received widespread government support in EU countries, which has resulted in the establishment of many national or international research organizations to facilitate research and development in this demanding area, which is so closely related to the environment and the very existence of the human civilisation.

References

Annual Institute Report of Central Institute for Research on Cotton Technology (2013–2014) In: Chattopadhyay SK (ed) Preface, P(i) (2014)

Avella M, De Vlieger JJ, Errico ME, Fischer S, Vacca P, Volpe MG (2005) Biodegradable starch/clay nanocomposite films for food packaging applications. Food Chem 93:467–474

Azeredo HMC (2009) Nanocomposites for food packaging applications. Food Res Int 42:1240–1253

Berthet MA, Angellier-Coussy H, Chea V, Guillard V, Gastaldi E, Gontard N (2015) Sustainable food packaging: valorising wheat straw fibres for tuning PHBV-based composites properties. Compos A 72:139–147

Chen Y, Liu C, Chang PR, Cao X, Anderson DP (2009) Bionanocomposites based on pea starch and cellulose nanowhiskers hydrolyzed from pea hull fibre: effect of hydrolysis time. Carbohydr Polym 76(4):607–615

Chen J, Zhang Y, Sun J (2011) An overview of the reducing principle of design of corrugated box used in goods packaging. Procedia Environ Sci 10:992–998

Chen W, Wang X, Tao Q, Wang J, Zheng Z, Wang X (2013) Lotus-like paper/paperboard packaging prepared with nano-modified overprint Varnish. Appl Surf Sci 266:319–325

Cherian BM, Leão AL, Souza SF, Costa LMM, Olyveira GM, Kottaisamy M, Nagarajan ER, Thomas S (2011) Cellulose nanocomposites with nanofibres isolated from pineapple leaf fibers for medical applications. Carbohydr Polym 86:1790–1798

Davis G, Song JH (2006) Biodegradable packaging based on raw materials from crops and their impact on waste management. Ind Crops Prod 23:147–161

Dogan N, McHugh TH (2007) Effects of microcrystalline cellulose on functional properties of hydroxyl propyl methyl cellulose microcomposite films. J Food Sci 72(1):E16–E22

El-Wakil NA, Hassan EA, Abou-Zeid RE, Dufresne A (2015) Development of wheat gluten/nanocellulose/titanium dioxide nanocomposites for active food packaging. Carbohydr Polym 124:337–346

Feichtinger M, Zitz U, Fric H, Kneifel W, Domig KJ (2015) An improved method for microbiological testing of paper-based laminates used in food packaging. Food Control 50:548–553

Galic K, Scetar M, Kurek M (2011) The benefits of processing and packaging. Trends Food Sci Technol 22:127–137

Gallstedt M, Hedenqvist MS (2006) Packaging-related mechanical and barrier properties of pulp–fiber–chitosan sheets. Carbohydr Polym 63:46–53

Guazzotti V, Limbo S, Piergiovanni L, Fengler R, Fiedler D, Gruber L (2015) A study into the potential barrier properties against mineral oils of starch-based coatings on paperboard for food packaging. Food Packag Shelf Life 3:9–18

Guillaume C, Pinte J, Gontard N, Gastaldi E (2010) Wheat gluten-coated papers for bio-based food packaging: Structure, surface and transfer properties. Food Res Int 43:1395–1401

Gurav SP, Bereznitski A, Heidweiller A, Kandachar PV (2003) Mechanical properties of paper-pulp packaging. Compos Sci Technol 63:1325–1334

Haafiz MKM, Hassan A, Zakaria Z, Inuwa IM (2014) Isolation and characterization of cellulose nanowhiskers from oilpalm biomass microcrystalline cellulose. Carbohydr Polym 103:119–125

Hakovirta M, Aksoy B, Hakovirta J (2015) Self-assembled micro-structured sensors for food safety in paper based food packaging. Mater Sci Eng, C. doi:10.1016/j.msec.2015.04.020

Indian Standard, IS: 7406 (Part 1) (1984) Specification for jute bags for packing fertilizer, part 1: laminated bags manufactured from 407 g/m^2, 85 × 39 Tarpaulin fabric

Indian Standard, IS: 7406-2 (1986) Specification for jute bags for packing fertilizer, part 2: laminated bags manufactured from 380 g/m^2, 68 × 39 Tarpaulin fabric

Indian Standard (IS) 12626 (1989) Laminated jute bags for packing milk powder—specification

Indian Standard, IS 2566 (1993) Textiles—B-twill jute bags for packing foodgrains—specification

Indian Standard, IS 1943 (1995) Textiles—A-twill jute bags-specification

Indian Standard, IS 9685 (2002) Textiles—sand bags—specification

Indian Standard (IS) 12650 (2003) Textiles—jute bags for packing 50 kg foodgrains—specification (second revision)

Indian Standard (IS) 15138 (2010) Textiles—jute bags for packing 50 kg sugar—specification

Indian Standard, IS 12269 (2013) Ordinary Portland Cement, 53 grade—specification

Lavoine N, Givord C, Tabary N, Desloges I, Martel B, Bras J (2014) Elaboration of a new antibacterial bio-nano-material for food-packaging by synergistic action of cyclodextrin and microfibrillated cellulose. Innov Food Sci Emerging Technol 26:330–340

Li J, Wei X, Wang Q, Chen J, Chang G, Kong L, Su J, Liu Y (2012) Homogeneous isolation of nanocellulose from sugarcane bagasse by high pressure homogenization. Carbohydr Polym 90:1609–1613

Lim LT, Auras R, Rubino M (2008) Processing technologies for poly(lactic acid). Prog Polym Sci 33:820–852

Lu Y, Weng L, Zhang L (2004) Morphology and properties of soy protein isolate thermoplastics reinforced with chitin whiskers. Biomacromolecules 5:1046–1051

Mahalik NP, Nambiar AN (2010) Trends in food packaging and manufacturing systems and technology. Trends Food Sci Technol 21:117–128

Majeed K, Jawaid M, Hassan A, Bakar AA, Khalil HPSA, Salema AA, Inuwa I (2013) Potential materials for food packaging from nanoclay/natural fibres filled hybrid composites. Mater Des 46:391–410

Mensitieri G, Maio ED, Buonocore GG, Nedi I, Oliviero M, Sansone L, Iannace S (2011) Processing and shelf life issues of selected food packaging materials and structures from renewable resources. Trends Food Sci Technol 22:72–80

Morais JPS, Rosa MF, Filho MMS, Nascimento LD, Nascimento DM, Cassales AR (2013) Extraction and characterization of nanocellulose structures from raw cotton linter. Carbohydr Polym 91:229–235

Muthu SS, Li Y, Hu JY, Mok PY (2009) An exploratory comparative study on eco-impact of paper and plastic bags. JFBI 1(4):307–320

Muthu SS, Li Y, Hu JY, Mok PY (2011) Carbon footprint of shopping (grocery) bags in China, Hong Kong and India. Atmos Environ 45:469–475

National Agricultural Innovation Project (2012) Final Technical Report of National Agricultural Innovation Project on "Synthesis and characterization of nano-cellulose and its application in biodegradable polymer composite to enhance their performance", Central Institute for Research on Cotton Technology, Mumbai, Year 2012, pp 1–28

Nayak P, Swain S (2002) Plastics and pollution: biodegradable polymers. Popular Plast. Packaging 47(10):66–78

Neto WPF, Silverio HA, Dantas NO, Pasquini D (2013) Extraction and characterization of cellulose nanocrystals from agro-industrial residue—Soy hulls. Ind Crops Prod 42:480–488

Pathare PB, Opara UL (2014) Structural design of corrugated boxes for horticultural produce: a review. Biosyst Eng 125:128–140

Qian X, Shen J, Yu G and An X (2010) Influence of pulp fiber substrate on conductivity of polyaniline-coated paper prepared by in situ polymerization. Bioresource 5: 899–907

Rhim JW, Lee JH, Hong SI (2007) Increase in water resistance of paperboard by coating with poly(lactide). Packag Technol Sci 20:393–402

Rosa MF, Medeiros ES, Malmonge JA, Gregorski KS, Wood DF, Mattoso LHC, Glenn G, Orts WJ, Imam SH (2010) Cellulose nanowhiskers from coconut husk fibers: Effect of preparation conditions on their thermal and morphological behavior. Carbohydr Polym 81:83–92

Salehudin MH, Salleh E, Mamat SNH, Muhamad II (2014) Starch based active packaging film reinforced with empty fruit bunch (EFB) cellulose nanofiber. Procedia Chem 9:23–33

Santos RM, Neto WPF, Silvério HA, Martins DF, Dantas NO, Pasquini D (2013) Cellulose nanocrystals from pineapple leaf, a new approach for there use of this agro-waste. Ind Crops Prod 50:707–714

Savadekar NR, Mhaske ST (2012) Synthesis of nano cellulose fibers and effect on thermoplastics starch based films. Carbohydr Polym 89:146–151

Savadekara NR, Karandea VS, Vigneshwaran N, Bharimalla AK, Mhaske ST (2012) Preparation of nano cellulose fibers and its application in kappa-carrageenan based film. Int J Biol Macromol 51:1008–1013

Shah AA, Hasan F, Hameed A, Ahmed S (2008) Biological degradation of plastics: a comprehensive review. Biotechnol Adv 26:246–265

Silverio HA, Neto WPF, Dantas NO, Pasquini D (2013) Extraction and characterization of cellulose nanocrystals from corncob for application as reinforcing agent in nanocomposites. Ind Crops Prod 44:427–436

Siracusa V, Rocculi P, Romani S, Rosa MD (2008) Biodegradable polymers for food packaging: a review. Trends Food Sci Technol 19:634–643

Song YS, Begley T, Paquette K, Komolprasert V (2003) Effectiveness of polypropylene film as a barrier to migration from recycled paperboard packaging to fatty and high-moisture food. Food Addit Contam 20:875–883

Sorrentino A, Gorrasi G, Vittoria V (2007) Potential perspectives of bio nanocomposites for food packaging applications. Trends Food Sci Technol 18(2):84–95

Sozer N, Kokini JL (2009) Nanotechnology and its applications in the food sector. Trends Biotechnol 27(2):82–89

The Gazette of India, Extraordinary, Part II-Section 3-Sub-section (II), Ministry of Textile-Order dated 13th Feb 2015

Vigneshwaran N, Ammayappan L, Huang Q (2011) Effect of gum arabic on distribution behavior of nanocellulose fillers in starch film. DOI, Appl Nanosci. doi:10.1007/s13204-011-0020-5

Yoshino T, Isobe S, Maekawa T (2002) Influence of preparation conditions on the physical properties of Zein films. J Am Oil Chem Soc 79:345–349

Youssef AM, El-Samahy MA, Rehim MHA (2012) Preparation of conductive paper composites based on natural cellulosic fibers for packaging applications. Carbohydr Polym 89:1027–1032

Youssef AM, Kamel S, El-Samahy MA (2013) Morphological and antibacterial properties of modified paper by PS nanocomposites for packaging applications. Carbohydr Polym 98:1166–1172

Zhang Y, Chen J, Wu Y, Sun J (2011) Analysis on hazard factors of the use of corrugated carton in packaging low-temperature yogurt during logistics. Procedia Environ Sci 10:968–973

Environmental Impacts of Packaging Materials

Varun, Aashish Sharma and Himanshu Nautiyal

Abstract When humankind began to store food or other items for next-day use, packaging in its primitive form emerged. In ancient times, leaves and bushes were used. By using high-end lightweight, durable, and cheaper material, today's packing industry has evolved exponentially. The industry is continuously searching for packaging solutions that have better strength, are easier to handle, are hygienic, are lightweight, and, most importantly, are sustainable. The major packaging materials are plastic, polystyrene, cardboard, etc. All of these materials are low in cost, light in weight, and durable. The world's growing population has led to large amount of packaging waste, which further contributes to the problem of its disposal and other environmental issues. High-energy consumption (embodied) and environmental problems are associated with packaging materials, which underscores the need to regard the proper use of packaging materials from an environment point of view. To analyse and quantify the environmental impacts associated with various packaging materials, an effective methodology is required. Life-cycle assessment (LCA) is an effective tool that can be utilised to evaluate various environmental impacts of packaging materials. This chapter discusses the environmental impacts associated with packaging materials and the use of LCA to evaluate these impacts so that they can be reduced considerably.

Varun (✉)
Department of Mechanical Engineering, National Institute of Technology,
Hamirpur, Himachal Pradesh 177005, India
e-mail: varun7go@gmail.com

A. Sharma
Department of Mechanical Engineering, Lovely Professional University,
Phagwara, Punjab 144806, India

H. Nautiyal
THDC Institute of Hydropower Engineering and Technology,
Bhagirathipuram, Tehri, Uttarakhand 249124, India

S.S. Muthu (ed.), *Environmental Footprints of Packaging*,
Environmental Footprints and Eco-design of Products and Processes,
DOI 10.1007/978-981-287-913-4_5

Keywords Environmental impacts · Packaging · Sustainability · LCA

1 Introduction

As people's lifestyles have changed, so has the environment. There have been significant changes in our ecosystem. Currently many problems are associated with the disposal of waste. It can clearly be seen that during the growing phase of packaging, no attention was being paid to sustainable solutions to waste. Now it is responsibility of humankind to contribute toward a greener environment. As an initial step, some regulatory bodies have been set up that regulate pollution from different industries and force them to reduce emissions. In addition, a few companies are taking sustainability into consideration and are performing environmental analyses of their manufacturing techniques. This analysis is the most important process used in almost all areas of activities from that of a small needle to those of massive aircraft parts. Packaging has many applications in transportation, preservation, and storage, etc. and provides protection to goods from moisture, breakage, dust, and contamination, etc. The medical, food, and beverage industries are almost completely dependent on packaging. In ancient times, bags, boxes, cases, etc., were made of natural materials, which were commonly used for packaging; however, with the passage of time, more effective materials were developed that protect goods not only from contamination but also maintain the characteristics and properties of the goods. For example, beverages packed in sealed bottles have the same taste and effectiveness as when they were packaged.

As the packaging sector continues to grow, its mammoth contribution toward environmental hazards is continuously increasing. Products come to the consumer in different packages. Because the consumer market is growing, packaging waste is exceeding standards set by regulating bodies. Industry is moving toward greener solutions for packaging, and the amount of waste from packaging is expected to be stable by the year 2021 (www.transperancymarketresearch.com). Packaging materials differ depending on their use and include paper, plastic, metal, and glass. Examples of reusable packaging include drum sand (reusable steel drums for storing liquids i.e. oil etc), plastic containers, etc. On the basis of application, the sustainable packaging market can be bifurcated into food and beverage packaging, personal care packaging, appliance packaging, etc.

As the consumption of packaging materials has increased throughout the world, the problem of packaging waste and its disposal is now looming. Excessive use of packaging materials is creating many environmental problems. Large amounts of packaging material waste raises the requirement of effective waste-management systems. In addition, many packaging materials, e.g., polyethylene, are not suitable for disposal in the environment. However, some packaging materials can be recycled, but the environmental impacts associated with their manufacturing and transportation, as well as their disposal, leads to various environmental problems. Therefore, it is important to control the environmental impacts associated with packaging materials. The initial step in this task is to evaluate the environmental impacts of different packaging materials. A tool is required to evaluate these impacts in the form of various sustainability indicators. LCA is one of the most effective tools to measure and study the environmental impacts associated with a product's packaging. LCA helps us study the environmental impacts of different products during their entire life cycle. This chapter begins with the introduction and discussion of various packaging materials along with their environmental impacts. The main features of LCA as a methodology to evaluate environmental impacts, as well as its applicability to reduce the environmental impacts of packaging materials, are discussed. In addition, some case studies on LCA of packaging materials are presented.

2 Packaging Materials and Their Environmental Impacts

The availability of food and beverages is highly dependent on their packaging. Advances in packaging materials have made the preservation and transportation of food items possible around the world. The shelf life of products has increased with better packaging. In addition, demand for quality food has lead to packaging innovation, and these innovations in packaging have helped to create new food categories and added convenience (Risch 2009). The primary functionality of packaging is not limited to simply containing the product. With the development of different lifestyles, more people require quality foods that can be preserved for longer periods of time. Currently packaging has become multifunctional. It involves protecting the product from external gases; blocking light to protect foods and their nutrients, colour, and texture; and preserving the product by maintaining specific ambient conditions around the food inside a container (Risch 2009).

Original packaging materials consisting of natural materials, such as skins, bark, leaves, and woven twigs, worked marginally well because foods were preserved by drying, smoking, salting, or fermenting. Deficiencies in these materials led to the development of textile, wood, ceramic, and glass containers, although they also have limitations in protecting food adequately. The development of lithography in 1798 saw the rise of low-cost printing and the development of labels. Canned tomatoes were introduced around the time of American Civil War. Heat sterilisation of spoilable foods in metal and glass containers was introduced in the early nineteenth century. This was important step in the area of packaging. Following a century later was the development of frozen foods in paperboard packages that maintained the nutrition, taste, and convenience of perishables all year long. To protect pulverized tobacco from ambient moisture, metal cans were used during the period of the Industrial Revolution (James et al. 2005; Risch 2009). Later, Nicholas Appert developed the idea of using cans to preserve food for the French army. Glass bottles were then replaced by metal cans. The use of heat processing was increased when using metal cans compared with glass. In the 1890s, individual packaging was utilised for biscuits. Before this, biscuits were packaged in large containers, and customers were filled their bags with biscuits to take home. Liners inside the bags protected the biscuits from moisture. In the history of packaging, it was important step when customised packaging was invented for a product. For a tight seal of glass bottles, the metal cork was developed by William Painter in 1892. It reduced the influx of oxygen into the bottle. The packaging of food items was also influenced by the development of how customers shopped for those items (Verghese et al. 2011).

In the United States, the first supermarket came into existence in 1920. Essential requirement for the development of packaging and stores was goods in packages at that time. In New York, the concept of the "economy store" was introduced in 1907, and it was commercially successful. Due to this success, the first supermarket—named Piggly Wiggly—was opened in Memphis, Tennessee, USA, in 1916. In this type of store, customers could purchase items that were stored on shelves in aisles. In Houston, another company provided trolleys (shopping carts) to customers (Lewis 2011). The development of new distribution and packaging techniques increased during World War II. These developments include plastic films and thin metal foils and sheets. The most frequently used packaging material midway through the nineteenth century was polyethylene. The manufacturing of ethylene packaging material was patented by Imperial Chemical Industries. The process involved compressing ethylene gas and heating it to a high temperature. During the mid–twentieth century, single-use packaging containers were introduced into the marketplace to replace refillable containers to some extent. The dynamics of the distribution chain were changed by this development (Verghese et al. 2011).

Today a variety of packaging materials, such as bottles, cellophane, cartons, plastics, cans, etc., are available; however, with the development of new packaging materials, the problem of their waste management has also increased. The consumption of packaging materials is increasing drastically in almost all nations of the world. Many environmental impacts are associated with the production, operations, transportation, and disposal of packaging materials. Packaging industries play a large role in the contamination of land, air, soil, and water. Therefore, it becomes important to analyze the environmental impacts of packaging materials in terms of moving toward a sustainable future. This is due to the fact that packaging materials cannot be removed from daily life because they have become an important part of all areas of human activities; however, their environmental impacts can be controlled considerably. This can be done by evaluating the environmental impacts of packaging materials using effective methodologies that evaluate environmental indicators in quantitative terms.

3 Life Cycle Assessment (LCA) and Sustainability

LCA is an effective tool used to evaluate the environmental effects associated with a product, process, or service during its entire life cycle i.e., "cradle to grave." The concept of LCA was introduced in the early 1880s by an economist, Patrick

Geddes, who proposed efficiency improvements to the product life cycle of coal as an energy source (IPCC 2001). At an early stage, the focus of life-cycle analysis was on energy balance, technology, and society's dependency on alternate energy sources such as nuclear energy (Hulme et al. 2002). Later, LCA methodology became standardised and rapidly developed as a practice within the ISO 14040 environment standard series (Sartori and Hestnes 2007; Sayal et al. 2006). Now LCA is seen as a tool that measures the variable inputs and outputs of any consumer product, buildings, packaging, etc. One of the largest achievements of LCA is becoming a part of the sustainable decision-making process in multinational companies such as Toyota (Energy Information Administration 1997). The initial step in such companies is to develop a sustainable corporate strategy that clarifies specific business cases for sustainable development. Determining the environmental life-cycle impacts of the company's products and services will lead toward more sustainable products and services. A recent development has been seen in the life-cycle initiative wherein the United Nations Environment Programme (UNEP) and the Society for Environmental Toxicology and Chemistry (SETAC) collaborated to develop an understanding and practice of life-cycle thinking (California Energy Commission 1998).

The term "sustainable development" came into being when the World Commission on Environment and Development published its landmark report, *Our Common Future*, in 1987. "Sustainability" can be defined as development that meets current demands without compromising the ability of future generations to meet their demands. This term represents the environmental, economic, and social balance of products and services. In 1997, John Elkington (United States Department of Energy 1999) popularized the term "triple bottom line (TBL)." TBL can be explained as a concern for "people, profit' and planet." TBL became a term to represent a common ground for sustainability following the debate over sustainable development, wherein economics must be balanced with the current and future needs of society and the environment (Arena and De Rosa 2003).

3.1 LCA Methodology

Life-cycle assessment (LCA) is a methodology that acts as a tool using qualitative assessment of materials, energy flows, and environmental impacts associated with materials and processes. It is used for systematically estimating the environmental impact of each material and process. LCA is a technique used for evaluating different parameters and aspects, e.g., greenhouse gas (GHG) emissions, associated with the fabrication of a product and its impacts throughout the lifetime (i.e., cradle to grave) of a product, e.g., extraction of raw materials, manufacturing, consumption, and disposal (Kim 1998). Life-cycle analyses are also associated with the evaluation of energy and materials used and waste material discharged into the environment during the product's life cycle. The technical framework of LCA consists of four key components that play an important role in assessment. They

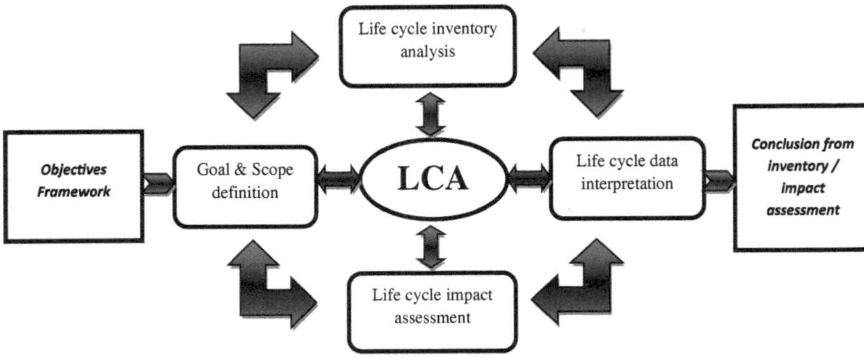

Fig. 1 Stages of life cycle assessment (Sharma et al. 2011)

are interrelated throughout the process and in accordance with the terminology of ISO (International Standards Organization). LCA methodology consists of four stages: goal and scope definition, life-cycle inventory (LCI) analysis, life-cycle impact assessment, and life-cycle interpretation (Fig. 1).

Definition of the LCA goal and scope establishes the functional unit usually focusing on the most important impact categories, system boundaries, and quality criteria for the inventory data. Common examples of LCA environmental indicators are shown in Fig. 2.

LCI analysis is associated with the accumulation and processing of data on materials and energy flows during various stages of the product's life cycle. In life-cycle impact assessment (LCIA), the environmental impacts of various flows of material and energy are assigned to different categories of environmental impact. Finally, life-cycle interpretation involves the interpretation of results from both LCI analysis and life-cycle impact assessment. It includes the identification of significant issues by pinpointing them as well as the evaluation of results, which are based on the data collected. The ISO has defined LCA as "a compilation and evaluation of the inputs, outputs and the potential environmental impacts of a product system throughout its life cycle" (Fig. 3).

3.2 Strength of LCA

LCA considers all environmental hazards caused by the release of emissions to any environmental compartment. It starts from the extraction of raw materials and energy used to manufacture a product through the product's consumption or use phase to final disposal of the product. LCA encourages companies to take a better approach toward protecting the environment by choosing better methods or processes for product development. LCA acts as an "alarm" by highlighting types of environmental hazards in developing a product or process and at which stage of

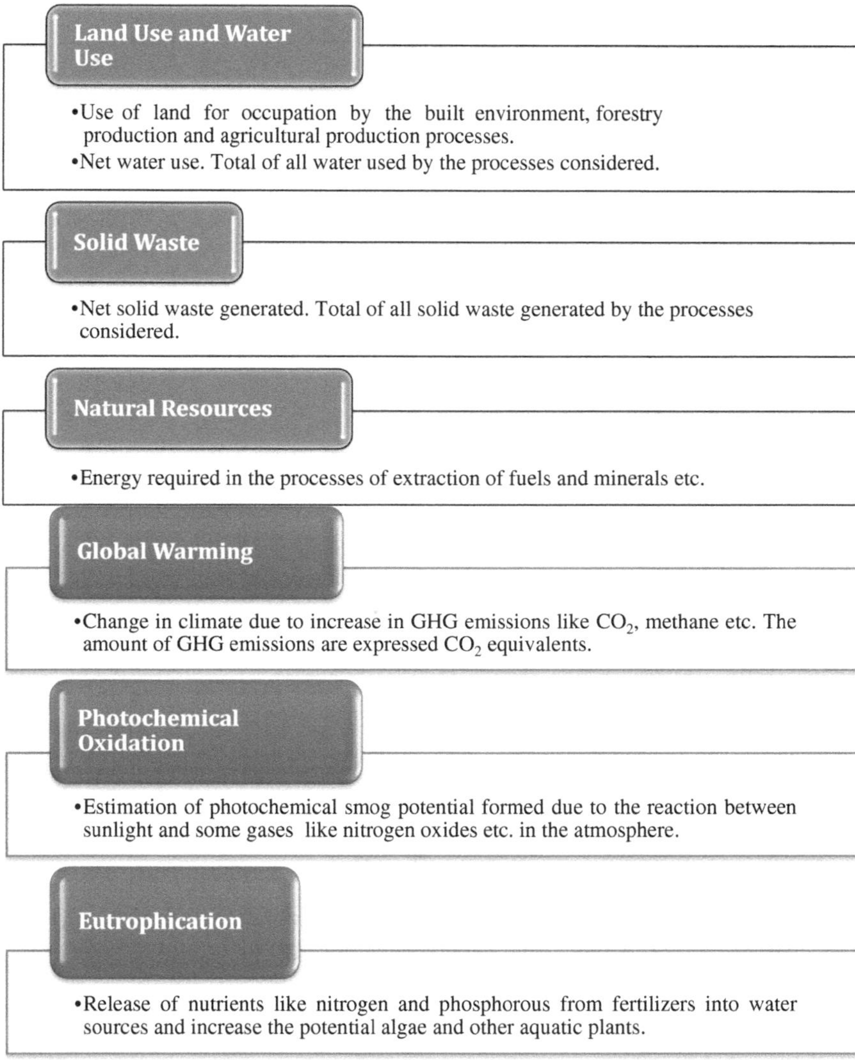

Fig. 2 Some common LCA environmental indicators

production, use, or disposal the hazard(s) will most affect the environment. LCA prioritizes products or processes that are highly likely to deteriorate the environment, and promotes ways to change that item or process to decrease its environmental impact.

Sometimes LCA can also be used as a tool for the decision maker when he or she is seeking a better alternative by comparing all of the environmental impacts caused by the production of a product. The results of LCA can help the person to choose the optimum production process as well as benefit the company from

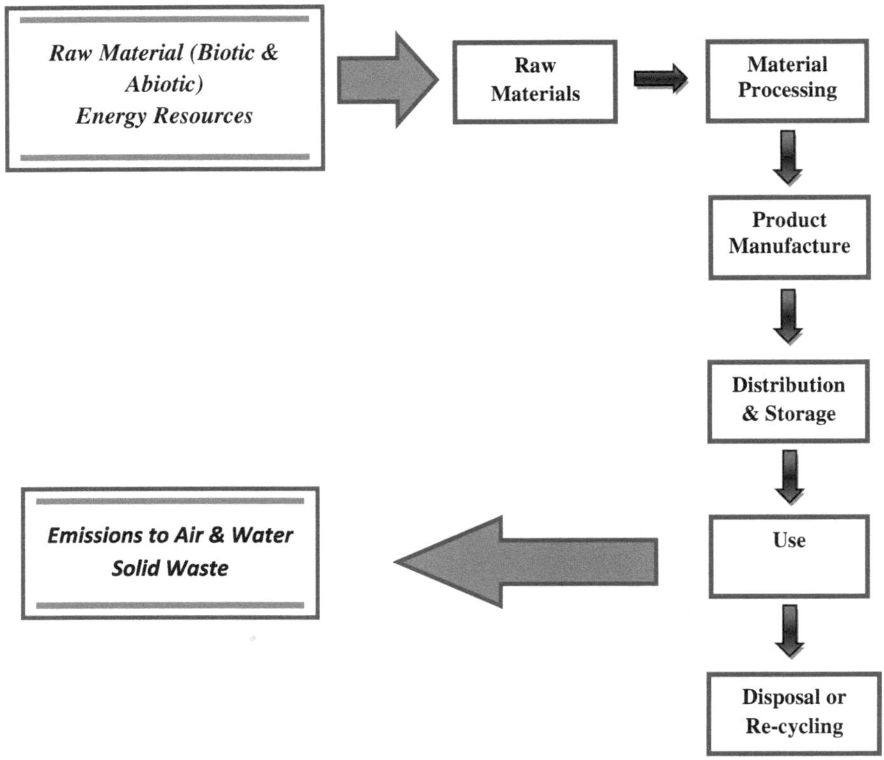

Fig. 3 Life-cycle system concept (Verghese 2008)

economic and social points of view. Although LCA has its own benefits, sometimes it can also mislead if the database is compromised; therefore, there is a need for transparency in LCA modelling (Finnveden et al. 2009).

3.3 Potential Gaps

LCA is not an easy task to perform; rather, it requires many resources, and in some cases it may consume a lot of time. If the user wants to examine all of the information in detail, data compilation may be an issue. As mentioned previously, compromising the database can lead to inaccurate or misleading results. Therefore, before performing LCA, the availability of a data source, financial resource, and time for completing the analysis must be determined. The role of LCA is to provide correct information to the user so that environmental impacts associated with the health of our environmental surroundings are correctly estimated. However, it does not account for the performance and social acceptance of the product. It is

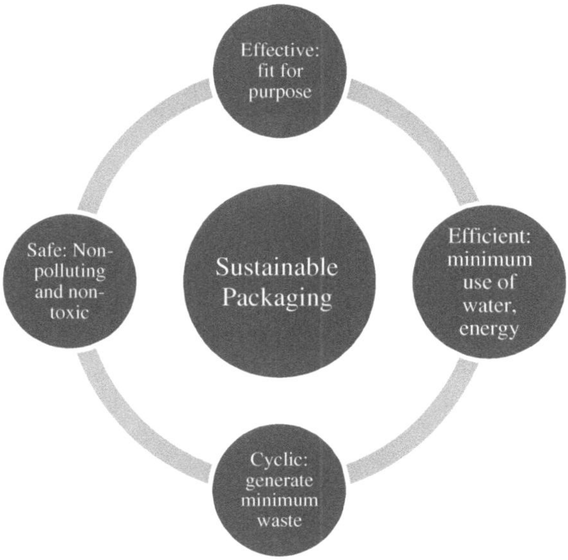

Fig. 4 Sustainable Packaging Alliance framework for packaging sustainability

not a cost-estimating analysis; hence, these factors must be assessed separately or accounted for within the company. The final decision is always made by the user; hence, the results of LCA help a decision maker make fact-based decisions.

For enhancing definitions, many sustainable packaging frameworks have been developed globally. The Australia-based Sustainable Packaging Alliance (SPA) in 2002 proposed that sustainable packaging must consider the following (Lewis 2002):

- Entire product life cycle;
- Triple bottom line (TBL); and
- Minimise environmental impacts of packaging

SPA has optimised their approach over time by adding a series of key performance indicators (KPIs) to four strong pillars of a framework for sustainable packaging (Fig. 4) (Lewis 2011).

4 Life-Cycle Assessment (LCA) in the Packaging Industry

A considerable number of LCAs were performed during the 1980s and 1990s on food packaging due to new packaging formats. For packaging of carbonated beverages, aluminium cans were used after tin-plated steel cans in the 1950s. A can opener was used for opening metal-can packaging. After 1963, the ring pull was introduced, and the stay tab was introduced in 1975. In recent times, application of modified-atmosphere packaging has increased. Due to such types of packaging, the shelf life of the product has lengthened. The speed of oxidation decreases when using gases, such as nitrogen or carbon dioxide, and reduces the growth of aerobic bacteria growth. Compared with bulk packaging, single-serve packaging has grown more in recent years. Single-serve packaging does not contain food in bulk quantities; only a defined quantity of food or beverage is contained by such packaging.

There exists a balancing act between appropriate serving size and changes in demographics and lifestyles. In Western society, as households become smaller and working hours outside the home increase, manufacturers are introducing smaller serving sizes and ready-to-go meals. As lifestyles continue to change, packaging is also trying to meet the challenge of delivering product. One must not forget that this rapid demand for packaging comes with environmental consequences that must be acknowledged, managed, and balanced (Verghese 2008).

4.1 Case Studies

During the last five decades, one of the main applications of LCA has been in the food- and beverage-packaging industry. For the packaging of food and beverages, the most commonly used packaging materials used today are glass, tin cans, and plastic. In 1969, Coca Cola was the first company to undertake such a study; at the time it was known as "resource and environmental profile analysis" (REPA). This was at the time when single-use packaging containers were being introduced to the market, and Coca Cola was interested in knowing the environmental profile of such types of packaging material compared with refillable containers. Since then, LCAs have been undertaken on many different packaging formats across the world to better understand the dynamics of materials selection, inform the design of packaging formats, and argue for better waste-management practices of used packaging (Lewis 2011). Packaging is designed in the most efficient manner to serve its purpose. Various LCA studies were performed to understand the impacts of packaging on the environment by considering it as a part of a product's life cycle. In most cases, approximately 2–5 % of overall environmental impacts are from packaging in the case of foods and 25 % in the case of beverages (Verghese et al. 2013).

Numerous packaging materials are available globally for food and beverage packaging. A study performed by Huang and Ma (2004) showed that the most common and popular packaging materials are polyethylene terephthalate (PET) containers, high-density polyethylene (HDPE), polypropylene (PP), polystyrene (PS), steel containers, aluminum containers, glass, cardboard boxes, liquid paperboards (LBP), etc. The study was performed using an integrated approach involving a combination of qualitative and quantitative methodologies. In an LCA approach, which is quantitative, aluminum and glass containers are considered to be less environment friendly because of the carcinogen and heavy metals emitted during their production; however, in a qualitative approach, the results are the opposite (Huang and Ma 2004). Therefore, a material cannot be judged as better compared with another material until the former performs well in all of the environmental measures, i.e., from packaging production, including raw material extraction, to the end its life.

According to Sebastien et al. (2009), in a comparison was performed between two baby food alternatives, i.e., glass jar versus plastic pots, plastic pots were deemed more beneficial than glass jars (Sebastien et al. 2009). The methodologies included in the study were IMPACT 2002+ and CML 2001. The environmental impacts reduced with the use of plastic pots instead of glass jars were 14–27 % of energy, 28–31 % for global warming potential, 31–34 % for respiratory inorganics, and 28–31 % for terrestrial acidification/nitrification. These benefits are due to production process, the light weight of the plastic, and the preservation process.

An LCA was performed on beer to calculate the environmental impacts and to determine possible betterment in the production and distribution phases. The stages included were the agriculture phase to the product delivery phase, but the consumption phase was excluded. The functional unit used was 505 multipacks of bottled beer. For calculation of KCL-ECO and for impact assessment, the DAIA 1998 method was used in the life-cycle assessment phase. Raw-water treatment, energy production, and contribution of oxygen depletion, eutrophication, summer smog, climate change, and acidification were also taken into consideration. In terms of both economic and environmental benefits, it was further recommended to optimize transport and to determine ways to diminish waste generation and save electricity.

As previously discussed, one of the major problems associated with packaging is waste management. After their use, many packaging materials end up in landfills. This problem can only be solved if the packaging material decays or decomposes along with the food item stored in it so that both can be disposed of together. The preferred option in the UK for dealing with organic waste is composting (www.defra.gov.uk). Composting is an environmentally responsible waste-management option involving the biodegradation of organic materials under aerobic conditions (Song et al. 2009).

Biodegradable polymers can be developed from renewable or nonrenewable (fossil) energy resources (Scott 2000). In a United States–based study performed in a school of packaging, the Michigan State University assessed the environmental profile of PET (polyethylene terephthalate), PS (polystyrene), and PLA (polylactic acid) containers. The methods used were ISO standards 14040, 14044, and 14049 as well as ASTM standard 7075. In this study, processes from corn-harvesting to corn-cracking were considered. The data of study were taken from Europe, North America, and the Middle East, and 1000 containers having a 1-pound capacity served as the functional unit. The environmental impacts measured were global warming, aquatic ecotoxicity, aquatic eutrophication, ozone depletion, aquatic acidification, respiratory organics and inorganics, nonrenewable energy, and land occupation. The results of the study showed that PLA (biodegradable) had a smaller environmental footprint than the petroleum-based PS, PP, and PET; among all choices, PET had the greatest impact value in terms of the respiratory inorganics, nonrenewable energy, and global-warming categories except for respiratory organics and aquatic acidification (Madival et al. 2009).

To know the complete performance of a packaging material, it is necessary to see the recycling phase of LCA. An Italian-based joint research team (CONAI) studied the Italian system of plastic packaging recycling and collected and mechanically recycled PET and PE liquid containers. The "basket-of-products" method is used for the comparison of resource consumption and environmental pollution by different management scenarios producing different products. It was concluded that the tool used is beneficial for the comparison of political waste-management scenarios. The amount of energy used for the production of 1 kg of recycled PE and PET is the lowest amongst all of the material compared in the study (Arena et al. 2003). However, other studies have given importance to the comparison of waste disposal with recycling (Craighill and Powell 1996). The methodology used is "life-cycle evaluation" (a combination of life cycle assessment plus economic evaluation). One tonne each of glass, paper, steel, aluminum, HDPE, PET, and PVC plastic waste were analysed separately, and comparison was made between the recycling system and its management by a waste-disposal system. It was found that the waste-disposal system contributed more to global warming than does recycling.

Toniolo et al. (2013) performed a study to determine to what degree the recycling of packaging is environmentally friendly. Comparison was made with the environmental effects of plastic food packages (multilayered plastic tray, PET tray), and how much the end-of-life treatment affects the environment was measured quantitatively. For multilayer film, the treatments are land-filling and incineration; for mono-material the treatments are recycling, land-filling, and incineration. The methods used were ReCiPe 2008, IMPACT 2002+ for impact assessment, CUT OFF approach for recycling, and Monte-Carlo technique for uncertainty analysis. Result shows that packaging by using recyclable material is preferable over the packaging material which is not recyclable.

In India, the All India Glass Manufacturers' Federation (AIGMF) engaged with PE sustainability solutions Pvt. Ltd. (PESSPL) to perform LCA of glass containers compared with alternative packaging (PET, aluminium can, carton, and pouch). The objectives of the study were (1) to evaluate the environmental footprint of glass containers compared with the alternatives and (2) to help AIGMF member companies project a "green image" of their product among consumers and stakeholders. CML 2001 method was used for the evaluation of environmental impacts. The functional unit was taken as 180 ml of liquor in a glass, PET, carton, or pouch container. The study found that emission, material inputs, and energy were affected over time by increasing the use of an abatement system, efficiencies, rebuilds, and cleaner technologies. It was recommended that for improvement in light-weighting, a "narrow-neck press-and-blow" technique should be used to package the liquor. In addition, importance was given to increasing the use of natural gas and renewable energy as well as the reuse of secondary materials (www.packagingconnections.com).

Table 1 Summary of examples of food and beverage packaging LCAs

Food/beverage item	Packaging type	Outputs
Retail packaging		
Milk	Glass HDPE, LLDPE, and PC pouches	Refillable HDPE preferred
Baby food	Glass jars versus plastic pots	
Coffee and butter	Flexible packing	Single-serving packaging; plastic preferred
Beer	Aluminium beverage cans	
Industrial packaging		
Large cartons	Reusable plastic pallets versus wooden pallets	Reusable plastic pallets have lower environmental impacts compared with wooden pallets

Detzel and Monckert (2009), Busser and Jungbluth (2009), Humbert et al. (2009), Madival et al. (2009), Falkenstein and Wellenreuther (2010), Keoleian and Spitzley (1999), Lee and Xu (2004)
HDPE high density polyethylene; *LLDPE* linear low density polyethylene

Most studies have been focused on the retail packaging level, whereas some have looked at industrial packaging (Table 1). There is no straightforward answer to which packaging format is best. The answer depends completely on geographical situation, functional, data quality, assumptions made, system boundaries selected, available waste-management practices, capture rates of materials, and the context of the situation.

The assessment of a packaging system is focused, but it does not consider the effect of foods and beverages. This is due to government and consumer concerns regarding environmental impacts of the actual packaging materials themselves without significant attention being paid to what is within the packaging (Williams et al. 2008). As the results of agricultural and food-production system studies have shown, the greater environmental impact of the product-packaging system involves the food or beverage contained within the packaging, not just the packaging materials themselves (Roy et al. 2009; Erlov et al. 2000; Jungbluth et al. 2000). For example, the overall resource efficiency of a coffee-packaging system is increased by packaging the coffee in a single-serving packet by decreasing material losses incurred during other coffee life-cycle stages such as production and use. Impacts associated with the production of the product itself are greater than the impacts related to the packaging production. A framework provided by LCA methodology can be utilised to measure the environmental impacts created by producing the product as well as the packaging (Busser and Jungbluth 2009).

Humbert et al. (2009) studied the comparative primary energy and greenhouse-gas emission impacts of glass jars versus plastic pots for baby food. Given the same mode of transportation, 14–27 % less primary energy is needed to transport plastic pots, and the production of plastic pots generates 28–31 % less global warming potential impact than the glass option. To influence the final impacts of the two alternatives, the actual material production, packaging weight, and on-site preservation parameters were identified. Keoleian and Spitzley (1999) compared HDPE, linear low-density polyethylene (LLDPE), glass paperboard carton, pouches, and polycarbonate packaging systems for the delivery of 1000 gallons of milk. Refillable HDPE, polycarbonate bottles, and flexible pouches were identified as preferable by the study. To understand the environmental impacts of three clamshell-packaging options made of PLA, PET, or PS, all three materials were evaluated by applying a life-cycle framework. PET was the least preferred option due to the greater weight of the containers. Resin production and transportation contributed to the environmental impact exerted by the packaging option (Madival et al. 2009).

Utilising multiple indicators as recommended by Roy et al. (2009) was used to compare the different environmental impacts of food options. All trade-offs associated with one specific mode of production compared with another could not be captured using a single indicator. For example, organic production is the preferred option when comparing conventional with organic agricultural practices. However, a more complete life-cycle study must also consider arable land use as a metric when comparing both agricultural practices because greater amounts of arable land are consumed by organic production to deliver the same service.

Shopping bags are also popular for the temporary packaging of items. Due to their ease of use, they are widely employed throughout the world. Muthu et al. (2011) performed a study to evaluate the carbon footprints of various shopping bags, e.g., plastic, paper, nonwoven, and woven type, using LCA technique with SIMAPRO 7.2. The study was performed for bag-users in Hong Kong, India, and China. The results showed that the carbon footprints of shopping bags are high in the absence of proper use and disposal options. In addition, the reuse of bags was found to be an important measure to significantly reduce carbon footprints. Another study was performed by Muthu et al. (2009) on plastic and paper bags, and their environmental impacts were compared using LCI data. Data on energy consumption and emissions during the manufacturing phase were used for LCI data. Plastic bags were found to be better than paper bags from an environmental point of view.

Muthu et al. (2013) has discussed a novel test instrument to estimate the eco-functional properties in terms of the reusability, impact strength, and weight holding–capacity of shopping bags. Result shows that paper bags were found better than plastic bags for single use category, whereas plastic bags were found superior in reusable category.

The energy flows associated with packaging are low when comparing the amount of energy invested in the various sectors involved in food production. According to previous investigations, total primary energy consumption is in the range of 7.3–10 units to produce 1 unit of food energy in the United States (Hall et al. 1986; Heller and Keoleian 2003). Manufacturing packaging material contributes 9 % of the total energy invested to produce food. Manufacturing packaging material consumes 1000 PJ out of the 11,000 PJ consumed yearly by the United States food sector (Pimentel and Pimentel 1996). In addition, the rapid conversion of prime farmland, the political problem of illegal workers, the depletion of topsoil, and the rate of groundwater withdrawal were identified as the key parameters posing a significant risk to the long-term sustainability of the United States food sector. In food production, the impacts of manufacturing packaging materials are less than impacts contributed by other sectors such as transportation, processing, and agricultural production.

5 Improvement in Sustainability Packaging Using Life-Cycle Thinking

Numerous environment-evaluation tools help in making the quick decisions that can be required in selection, design, and packaging system formats. The development of tools extend from guidelines and paper-based checklists to Internet-accessible evaluation tools and interactive and life-cycle based analytical tools incorporating life-cycle methodology as a fundamental component (Verghese and Lockrey 2011). A summarized detail of available tools is tabulated in Table 2. A series of packaging objectives is presented in the table along with a tool that could be helpful in addressing the objective concerned. Information about the results obtained from each tool, along with a description on how to find more information on the particular tool, is briefly presented. An in-depth detail of tools describing their features, data sources, ease of use, types, timing of use, and rationales can be found in the available literature.

To render these tools effective, it is vital to have in place well-documented and -communicated processes that guarantees implementation. The important requirements to be considered for the selection and implementation of decision-support tools for a new product-development process are as follows (Verghese 2008):

- The tool must enable a simple work flow for the user by being instinctual, easy to communicate, and logical.
- The tool should fit into the company's culture.
- The tool should require the least amount of set-up time to make use of it.
- The tool should require less data-input requirements so as to make it user-friendly (i.e., the user should easily understand the benefits and features of the tool).
- The tool should present the results in a visually appropriate layout that is easily adoptable.
- The tool should include matters that relate to users on a day-to-day basis.

Table 2 Selection of packaging evaluation tool

Packaging objective	Suitable tool to use	Type of result
Obtain a general idea of the supply chain of a packaging system	Life-cycle map	Provides the design team with high-level scan of key materials and processes and allows the team to become familiar with potential environmental impacts across the supply chain
Design a new packaging system or update an existing design and begin an assessment of environmental impact and credentials of the design	Guidelines	Contains questions relevant to the materials and design of packaging and highlights particular areas, e.g., material selection and end-of-life issues, where additional information may be required
	Sustainable Packaging Alliance—packaging sustainability framework	Relates the four guiding principles of packaging sustainability (efficient, effective, cyclic, and safe) and identifies design strategies per principle and performance indicators (Sustainable Packaging Alliance 2010)
	Australian Sustainable Packaging Guidelines	Assessment of environmental impact and credential of packaging design and criteria (APCC 2009)
	Sustainable Packaging Coalitions–Design Guidelines For Sustainable Packaging	DfS background and strategies based on packaging environmental-impact credentials and criteria (Design guidelines for sustainable packaging 2006)
	WRAP–Guide to evolving packaging design	DfS data on consumer views, law, brand, innovation, tools, technique, and materials consideration; provides further links and glossary
	Envirowise–packaging design for environment	General packaging information, regularity data, material figures, and links to further reports (Envirowise 2008a)
	Envirowise–pack guide; a guide for packaging eco-design	Design-focused information on material regularity data and link to further resources (Envirowise 2008b)
Evaluate the environmental profile of several different packaging material combination	Packaging-specific analytical tool	Provides life cycle-based environmental information

(continued)

Table 2 (continued)

Packaging objective	Suitable tool to use	Type of result
Perform a quick scan of industry benchmark for packaging weight by material/formal		Provides industry average data on packaging-material weight for a range of packaging applications
Screen packaging design against alternatives	EDIT	Generates comparative graphical/tabular results for a range of environmental indicators for up to six different designs (Envirowise 2010)
	Wal-Mart–Package Modelling	Life cycle–based environmental impact metrics including comparative analysis and report exporting (Walmart 2008)
	Toyota–EPIC	Toyota packaging environmental indicator comparison, life-cycle cost comparison, life-cycle stage analysis, and life-cycle inventory
Screen packaging design against alternatives	Packaging-Specific LCA-based analytical tool	
	PIQET	Rapid streamlined environmental-impact assessment including comparative, tabulated graph- and inventory-based reporting; reports a range of packaging-sustainability metrics (Verghese et al. 2010)
	COMPASS	Reports include component contribution, comparative life-cycle assessments, and packaging attributes that are important to a designer (COMPASS 2009)
Model in detail the life-cycle environmental impacts of primary and alternative packaging systems and/or publish results	Life-Cycle Assessment Software	
	SimaPro	Detailed and flexible quantitative life-cycle environmental impact of a packaging system's full comparative and sensitive capability
	GaBi	Detailed and flexible quantitative life-cycle environmental impact of a packaging system's full comparative and sensitive capability

Certain things that should be taken into consideration include the following (Verghese and Lockrey 2011):

1. Who is the person in the organization using LCA as a tool?
2. The amount of detail required by the person performing LCA must be defined.
3. Why is LCA needed, and when is the right time to perform it?

5.1 Answering to Supplier Demands

Several suppliers are currently looking to apply sustainability initiatives through their supply chain, which often translates to food producers following new protocols and using new tools to attain these goals. Any place where environmental metrics are concerned, measurements are often underpinned by LCA data or methodology. Sustainable innovation products have an enhanced and improved environmental profile in which the enhancements are substantial and evident. To meet the requirements for sustainability, a product must show at least 10 % improvement throughout its life cycle in one of the crucial indicators—such as consumption of energy and/or water, total materials used for the product and its packaging, transport, or the application of renewable energy sources (instead of nonrenewable resources)—along with no noticeable worsening in any of the other indicators (White 2009).

In this definition, it is now a prerequisite now for supply chains to report a number of environmental indicators through the newly introduced supplier environmental sustainability scorecard. A strong use of life-cycle thinking creates environmental improvement involving "trade-off" decisions so as not to damage one indicator for the sake of other (e.g., focusing on carbon emissions only to see water and land use increase). As an additional sign of proactivity, Walmart has established a packaging scorecard and, more recently, a software platform called "Package Modelling," which permits supplied groups to be ranked and aggressively improved by modelling various improvements in design and increasing their rank in real time. These tools are also incorporated along with Walmart's already tested supply chain-management system, which exploits life-cycle data, both primary and generic, throughout the ranking process.

Private-sector supply chain alliances, e.g., those from Walmart and Proctor and Gamble, complement industry-based initiatives such as the Sustainable Packaging Coalition (SPC) in the US and the Sustainable Packaging Alliance (SPA) in Australia. These organizations have designed procedures and protocols that reinforce life-cycle thinking and deliver a platform for companies via the supply chain to vigorously cooperate in decreasing the environmental influence of food- and beverage-packaging products.

6 Future Scope

Future challenges exist related to application of LCA in the packaging industry, and they mostly concern low-carbon economy. It has been suggested by the Intergovernmental Panel on Climate Change (IPCC) that a 50–85 % reduction in CO_2-equivalent gases is required by 2050 to prevent a 2–2.4 °C increase in global temperature (IPCC 2007). Application of LCA in the packaging industry has several implications as follows:

1. increased accountability to customers regarding the ecological impact of packaging;
2. assisting packaging designers to reduce the ecological impact of packaging by shifting LCA from a reflective tool to an action-orientated decision-making tool; and
3. consideration to including food in up-scaling the functional unit within the packaging industry due to substantial emissions from food production.

6.1 Increased Demands from Consumers

There is a growing trend for businesses to be accountable for their actions under the banner of "corporate social responsibility." Sustainable packaging strategies have recently been introduced by Marks and Spencer and Walmart. Various indicators—e.g., $CO_{2eq.}$/tonne and innovation to meet their target of 5 % reduction in packaging across the supply chain—are addressed in Walmart's packaging scorecard. Different brands. such as Cadbury and Coca-Cola. have good CO_2–reduction targets in place. For example, to reduce 50 % of absolute carbon emissions by 2020 and reduce packaging used per tonne of product by 10 %, Cadbury's "purple goes green" commitment is in progress (www.cadburyinvestors.com). In addition, 92 % of retail markets in the UK are signatories of the Courtauld Commitment. This commitment aims to improve resource efficiency and reduce carbon emissions and the broader environmental impact of the retail grocery sector. These types of strategies require the packaging industry to be accountable for its ecological impacts, and LCA is well suited to measure such.

The packaging industry is facing increased consumer pressure to counter its image as the "visible face of waste" within the household. LCA plays an important role in communicating the worth of packaging in preserving and protecting the product. For example, due to advanced packaging solutions, there is only 2 % food waste in the supply chain in Europe as compared with 30 and 50 % food waste in developing countries (PricewaterhouseCoopers LLP 2010). To counter consumers' perceptions, the objective approach of LCA will be increasingly required. Voluntary carbon-labelling schemes have been trialled for food (Hogan and Thorpe 2009) in various countries as follows:

- United Kingdom (Carbon Trust)
- United States (Carbon Fund)
- Germany (Product Carbon Footprint pilot labelling scheme)
- Sweden (Climate Marking) and the European Union (carbon footprint measurement toolkit),
- Japan (30 companies have participated in a pilot scheme funded and coordinated by the Japanese Ministry of Economy, Trade and Industry)
- South Korea (CooL Label)
- Thailand (carbon label being developed by the Thailand Greenhouse Gas Management Organisation)

In the United Kingdom, consumers of products associated with CO_{2eq} emissions are informed by the United Kingdom's Carbon Trust label across the entire life cycle of the product. The Carbon Trust label consists of four "sub-labels": (1) the footprint, (2) the carbon footprint estimation expressed in CO_2-equivalent terms, (3) an endorsement by the Carbon Trust, and (4) a commitment by producer to minimise emissions (Hogan and Thorpe 2009). An educational component is included as an optional element to explain how the carbon footprint is calculated. This provides the packaging industry an opportunity to communicate the worth of the packaging. In allowing customers the advantage of knowing a product's carbon impact during its life cycle, the combination of sub-labels in the Carbon Trust label gives the consumer a full picture of the environmental consequences of purchasing, using, and disposing of a particular product (e.g., water use or human toxicity).

7 Conclusion

The consumption of packaging materials is increasing at a high rate throughout the world. Packaging materials are used in almost all areas of activities and have become a basic need in various industries. Many packaging materials can be recycled, but many environmental problems are associated with doing so. Thus, it is important to evaluate the environmental impacts of packaging materials throughout their entire life cycle and to considerably reduce their harmful effects on the environment. LCA as tool gives us the opportunity to identify the "grey areas" that affect our environment and thus indirectly affect us. Considering the effects of environmental issues associated with packaging materials, global investors now must develop alternate means of packing. Increased awareness toward environmental hazards is one of the major factors fuelling the global demand for green packaging. Due to this, considerable efforts are being made to decrease toxic waste and GHG emissions. Green packaging results in fewer toxic emissions and causes less pollution. Moreover, initiatives to clean up the environment, as well as strict regulations and monitoring agencies, are being enacted by governments globally.

References

APCC (Australian Packaging Covenant) (2009) A commitment by governments and industry to the sustainable design, use and recovery of packaging. Australian Packaging Covenant Council (APCC), Melbourne

Arena AP, De Rosa C (2003) Life cycle assessment of energy and environmental implications of the implementation of conservative technologies in school building in Mendoza-Argentina. Build Environ 38:359–368

Arena U, Mastellone ML, Perugini F (2003) Life cycle assessment of a plastic packaging recycling system. Int J Life Cycle Assess 8(4):92–98

Busser S, Jungbluth N (2009) The role of flexible packaging in the life cycle of coffee and butter. Int J Life Cycle Assess 14:80–91

California Energy Commission (CEC) (1998) Baseline energy outlook. P300-98-012, Aug 1998

COMPASS—Comparative Packaging Assessment (2009), www.sustainablepackaging.org

Craighill AL, Powell JC (1996) Lifecycle assessment and economic evaluation of recycling

Detzel A, Monckert J (2009) Environmental evaluation of aluminium cans for beverages in the German context. Int J Life Cycle Assess 14:70–79

Energy Information Administration (EIA) (1997) Commercial buildings characteristics 1995. US Department of Energy, Aug 1997

Envirowise (2008a) Packaging design for the environment: reducing costs and quantities. Envirowise, Oxfordshire

Envirowise (2008b) Packguide: a guide to packaging eco-design. Envirowise, Oxfordshire

Envirowise (2010) EDIT—The eco-design indicator tool. Envirowise, Oxfordshire

Erlov L, Lofgren C, Soras A (2000) PACKAGING—a tool for the prevention of environmental impact. Packforsk, Kista

Falkenstein EV, Wellenreuther F (2010) LCA studies comparing beverage cartons and alternative packaging: can overall conclusions be drawn? Int J Life Cycle Assess 15(9):938–945

Finnveden G, Hauschild MZ, Ekvall T, Guinée J, Heijungs R, Hellweg S, Koehler A, Pennington D, Suh S (2009) Recent developments in life cycle assessment. J Environ Manage 91(1):1–21

Hall CAS, Cleveland CJ, Kaufmann R (1986) Energy and resource quality. Wiley Interscience, New York

Heller MC, Keoleian GA (2003) Assessing the sustainability of the US food system: a life cycle perspective. Agric Syst 76:1007–1041

Hogan L, Thorpe S (2009) Issues in food miles and carbon labelling. ABARE Research report 09.18. ABARE, Canberra

Huang CC, Ma HW (2004) A multidimensional environmental evaluation of packaging materials. Sci Total Environ 324:161–172

Hulme M, Jenkins G, Lu X, Turnpenny J, Mitchell T, Jones R, Lowe J, Murphy J, Hassell D, Boorman P, McDonald R, Hills S (2002) Climate change scenarios for the United Kingdom: the UKCIP02 scientific report. Tyndall Centre for Climate Change Research, University of East Anglia, Norwich

Humbert S, Rossi V, Margni M, Jolliet O, Loerincik Y (2009) Life cycle assessment of two baby food packaging alternatives: glass jars vs plastic pots. Int J Life Cycle Assess 14:96–106

International Panel on Climate Change (2001) Climate change: the scientific basis. Cambridge University Press, Cambridge

IPCC (2007) Summary for policymakers, in climate change 2007: mitigation. Contribution to Working Group III to the Fourth Assessment Report of the Intergovernmental Panel on Climate Change. Cambridge University Press, Cambridge

James K, Fitzpatrick L, Lewis H, Sonneveld K (2005) Sustainable packaging systems development. In: Leal Filho W (ed) Handbook of sustainability research. Peter Lang Scientific Publishing, Frankfurt

Jungbluth N, Tietje O, Scholz RW (2000) Food purchases: impacts from the consumer's point of view investigated with a modular LCA. Int J Life Cycle Assess 5(3):134–142

Keoleian GA, Spitzley D (1999) Guidance for improving life cycle design and management of milk packaging. J Ind Ecol 3(1):111–126
Kim S (1998) Life-cycle assessment of embodied energy for building materials-focused on high-rise apartments. In: Proceedings of the World Renewable Energy Congress (WREC), Florence, pp 1559–1562
Lee SG, Xu X (2004) A simplified life cycle assessment of re-usable and single-use bulk transit packaging. Packag Technol Sci 17:67–83
Lewis H (2011) Designing for sustainability. In: Verghese K, Lewis H, Fitzpatrick L (eds) Packaging for sustainability. Springer, London
Lewis H, Sonneveld K, Fitzpatrick L, Nicol R (2002) Towards sustainable packaging. Discussion Paper
Madival S, Auras R, Singh SP, Narayan R (2009) Assessment of the environmental profile of PLA, PET and PS clamshell containers using LCA methodology. J Clean Prod 17:1183–1194
Muthu SS, Li Y, Hu JY, Mok PY (2009) An exploratory comparative study on eco-impact of paper and plastic bags. J Fiber Bioeng Inform 1(4):307–320
Muthu SS, Li Y, Hu JY, Mok PY (2011) Carbon footprint of shopping (grocery) bags in China, Hong Kong and India. Atmos Environ 45(2):469–475
Muthu SS, Li Y, Hu JY, Mok PY, Mao YF, Li QH, Wu XX (2013) Assessment of eco functional properties of shopping bags: development of a novel eco-functional tester. Int J Clothing Sci Technol 25(3):208–225
Pimentel D, Pimentel M (1996) Energy use in fruit, vegetavle and forage production. In: Pimentel D, Pimentel M (eds) Food, energy and society. University Press of Colorado, Niwot, pp 131–147
Pricewaterhousecoopers LLP (2010) Sustainable packaging: threat or opportunity? PricewaterhouseCoopers LLP, London
Risch SJ (2009) Food packaging history and innovations. J Agric Food Chem 57:8089–8092
Roy P, Nei D, Orikasa T, Xu Q, Okadome H, Nakamura N, Shiina T (2009) A review of life cycle assessment (LCA) on some food products. J Food Eng 90:1–10
Sartori I, Hestnes AG (2007) Energy use in the life cycle of conventional and low-energy building: a review article. Energy Build 39:249–257
Sayal M, Hastak M, Mullens M, Sweaney A (2006) United States-India collaborative research directions in urban housing and supporting infrastructure. J Archit Eng 12(4):163–167
Scott G (2000) Green polymers. Polym Degrad Stab 68:1–7
Sebastien H, Rossi V, Margni M, Jolliet O, Loerincik Y (2009) Life cycle assessment of two baby food packaging alternatives: glass jars vs plastic pots. Int J Life Cycle Assess 14:96–106
Sharma A, Saxena A, Sethi M, Shree V, Varun (2011) Life cycle assessment of a building: a review. Renew Sustain Energy Rev 15:871–875
Song JH, Murphy RJ, Narayan R, Davies GBH (2009) Biodegradable and compostable alternatives to conventional plastics. Philos Trans Royal Soc B 364:2127–2139
Sustainable Packaging Alliance (2010) Rmit University, Helen Lewis Research, Birubi Innovation, Victoria University. SPA packaging sustainability framework. Sustainable Packaging Alliance, Melbourne
Sustainable Packaging Coalition (2006) Design guidelines for sustainable packaging, version 1. GreenBlue Institute, Charlottesville
Talve S (2001) Life cycle assessment of a basic lager beer. Department of Environmental Engineering, Tallinn Technical University
Toniolo S, Mazzi A, Niero M, Zuliani F, Scipioni A (2013) Comparative LCA to evaluate how much recycling is environmentally favorable for food packaging. Resour Conserv Recycl 77:61–68
US Department of Energy (1999) A. D. Little guide for evaluation of energy savings potential. Prepared for the office of Building Technology, State and Community Programs (BTS). US Department of Energy

Verghese K (2008) Environmental assessment of food packaging and advanced methods for choosing the correct material. In: Chiellini E (ed) Environmentally compatible food packaging. Woodhead Publishing, Cambridge, p 2008
Verghese K, Lockrey S (2011) Selecting and applying tools. In: Verghese K, Lewis H, Fitzpatrick L (eds) Packaging for sustainability. Springer, London
Verghese K, Horne R, Carre A (2010) PIQET: the design and development of an online 'streamlined' LCA tool for sustainable packaging design decision support. Int J Life Cycle Assess 15(6):608–620
Verghese K, Lewis H, Fitzpatrick L (eds) (2011) Packaging for sustainability. Springer, London
Verghese K, Lewis H, Lockrey S, Williams H (2013) The role of packaging in minimising food waste in the supply chain of the future. Report; CHEP Australia; Issue: 3.0
Walmart (2008) Sustainability Scorecard FAQs
White P (2009) Building a sustainability strategy into the business. Corp Gov 4:386–394
Williams H, Wikstrom F, Lofgren MA (2008) Life cycle perspective on environmental effects on customer focused packaging development. J Clean Prod 16:853–859
www.cadburyinvestors.com Stitzer, T. Corporate Responsibility and Sustainability
www.defra.gov.uk. DEFRA, 2006 .Food industry sustainability strategy
www.packagingconnections.com. Life cycle assessment of "Container Glass and Comparison with Alternate Packaging Mediums"
www.transperancymarketreserach.com. Report: Green Packaging Market-Global Industry Analysis. Trend, Size, Share and Forecast 2015–2021

Bibliography

Boylston S (2009) Designing Sustainable Packaging. Laurence King Publishing, London
Chiellini E (2008) Environmentally compatible food packaging. Woodhead Publishing, Cambridge
Jedlicka W (2009) Packaging sustainability. Tools, systems and strategies for innovative package design. Wiley, Hoboken
Sterling S (2007) Field guide to sustainable packaging. Summit Publishing, Chicago
Sustainable Packaging Coalition (2006) Design guidelines for sustainable packaging, version 1. GreenBlue Institute, Charlottesville

Bioprocessing of Metals from Packaging Wastes

Meenu Gautam, Divya Pandey and Madhoolika Agrawal

Abstract Packaging refers to the covering used to protect the product inside. Metals—such as iron, copper, and their alloys, i.e., brass and bronze, have been used for the packaging and storage of goods since ancient times. Unique properties of metals, particularly the ease of fabrication, strength, thermal and electrical conductivities, and ability to hold diverse materials securely in different states, make them an essential packaging material either as such or as composites with materials such as polymers, fibers, plastics, and ceramics. Boxes, cans, cylinders, and foils made from iron, aluminum, tin, copper, etc., are the most common and everyday examples of metal-based packaging; however, specialized packaging requirements, e.g., for electronic parts, composites based on different metals are preferred. After its end use, discarded packaging becomes a major contributor to waste generation. Completely metal-based packages can be recycled; however, this becomes expensive for composites. In such cases, landfilling is the most common disposal method, which may cause adverse impacts on human health through the contamination of groundwater and soil. This calls for effective and better alternate metal waste-management options that can help metal recycling and recovery. In this chapter, we present a brief introduction of metal-based packaging, their various methods of disposal, and recovery and recycling options with particular focus on biotechnological approaches. With the help of different examples and recent developments in the recovery and reuse of waste metals, potential sustainable and cost-effective solutions in managing metallic or metal-based packaging waste are discussed.

Keywords Metals · Composites · Recycling · Packaging waste

M. Gautam · D. Pandey · M. Agrawal (✉)
Laboratory of Air Pollution and Global Climate Change, Department of Botany, Banaras Hindu University, Varanasi, Uttar Pradesh 221005, India
e-mail: madhoo.agrawal@gmail.com

S.S. Muthu (ed.), *Environmental Footprints of Packaging*,
Environmental Footprints and Eco-design of Products and Processes,
DOI 10.1007/978-981-287-913-4_6

1 Introduction

Packaging is an essential part of our daily lives and is associated with all consumer products, transportation of goods, and services. Moreover, packaging reflects cultures, traditions, and lifestyles. In industrial goods, packaging is a technical and crucial component because it is not merely a covering but a protection, and many times it may have functional roles. Metals have always been an essential packaging material due to their strength, their ability to easily contain solids as well as liquids, fine finishing and crafting options, and durability. However, current packaging materials have become extremely diverse and specialized for packaging purposes. Modern packaging materials also contain metals as their key components specifically to impart properties of electrical and thermal conductivities, malleability, ductility, and mechanical strengths such as coordination polymers, plastic-like metallized polyethylene terephthalate, and metal-matrix composites (MMCs).

After removal of a product from its packaging, the packaging is usually rendered useless and ends up as a waste. For instance, packing materials from households constitute nearly 30 % of total municipal solid wastes (MSW). However, packing materials generated at domestic levels are largely recyclable; in fact, 48.5 % of it is recycled in United States. The recycling rates of metal-based packaging waste, e.g., steel, aluminum, glass, plastic, paper, and paperboard, were approximately 69.0, 35.8, 33.4, 13.5, and 71.3 %, respectively (US-EPA 2010). Because the production of goods, their transportation, and their consumption are increasing, there has been a surge in discarded packaging, the proper disposal, recycling, and reuse of which poses a huge challenge. In this scenario, recovering metals from such discarded materials becomes crucial because they are limited resources, and they contaminate the environment.

In this chapter, we provide an overview of different types of metal-based packaging, the evolution of such packaging, and the emergence of new materials that contain metals in different proportions as their constituents. Then the chapter discusses the issue of waste generation due to these packaging materials, the implications for the environment, and the practices adopted for managing such waste. The focal point is to present the prevalent and potential reuse and recycling techniques for metal-based packing waste. The chapter concludes with an outlook for the sustainable management of metal-based packaging materials in light of the ongoing progress in this direction.

2 Types of Metal-Based Packaging Materials

Metals have been in use since ancient times, and they were also some of the first materials used for the storage and transportation of goods in the form of boxes and containers. Precious metals, such as silver, were being used by royal families, whereas brass, aluminum, iron, and tin offered much cheaper alternatives for common use and also helped with the long-term storage and long-distance transportation of materials. In time, cheaper metals, stronger and lighter alloys, thinner foils and gauges, and versatile metal composites with polymers and other materials have evolved, which serve different packaging requirements. In addition to being used in their original form, metals also impart color, strength, and conducting properties, which make them an essential constituent of many other packing materials such as glass, polymers, plastics, papers, paperboards, etc. The most common metals used are chromium (Cr), lead (Pb), arsenic (As), cadmium (Cd), aluminum (Al), etc. Most of these metals become toxic to plants, animals, and humans when present in high concentrations in the environment.

2.1 Metals and Alloys

Metallurgy was invented in ancient times, and metals have become an important packaging material since then because they offer a multitude of design options, strength, and versatility for reuse. Even in modern times, metals as such are an unrivalled premium packaging option. One of the most common type metal-based packaging items is to use it directly in the form of barrels, cans, and boxes. Even though metal containers contain all kinds of items, including liquids, it was not a convenient option until an easy method of opening them was invented. The can opener was invented in 1875 and made metal packaging a suitable option even for households. Thereafter, more improvements were made leading to pop-top and tear-tab can lids near 1950, and currently tear tapes and screw tops have been invented for small packaging (Hook and Heimlich 2011). Besides their use in their pure form, metals are increasingly been used in combination with other materials such as glass, wood, plastic, and polymers. They form different parts such as lids, frames, screws, springs, and decorations. Such types of packaging are particularly useful in the food industry and even for small-scale domestic storage. Metals are impermeable to air and water and hence greatly reduce the chances of contamination. This property therefore provides longer shelf life, tamper evidence, total protection against external damage, and safe containment of reactive items such as food, paints, medicines, etc. Working with metal containers is easy and efficient because there are minimal losses at all stages of filling, sealing, packing, distribution, and sale. Therefore, metals are the popular choice in the form of drink cans, food cans, aerosol containers, tubes, open trays, caps, foil containers, etc. In addition, they are fully recyclable. Some commonly used metals and alloys are detailed below.

2.1.1 Aluminum

Al is a silvery white, nontoxic metal that is commonly used for making cans, foils, and laminated paper or plastic sheets. It is one of the safest packaging options when it comes to contact storage of food items. At pH < 4.5, Al uptake from uncoated food contact materials made of pure Al is affected by the acidity of the food product and the solubility of the salt formed (Cutter 2002). Hence, pure Al is not a preferable packing option for such items. Foodstuff having higher salt concentrations inside the packaging material may also increase the migration of metals (Elinder and Sjogren 1986). Al is normally coated for packaging applications. Moreover, in a range of applications in many industries, Al containers are used for different processes of transformation of food such as refrigeration, freezing, cooking, preservation, modified atmosphere sterilization, and pasteurization (Holdsworth and Simpson 2007). To improve its properties, Al is often made into alloys which are resistant toward corrosion. Al alloys may contain magnesium (Mg), silicon (Si), iron (Fe), manganese (Mn), copper (Cu), and zinc (Zn) (CEN 2004). Its compounds, particularly aluminum oxide, are used as coatings creating a barrier against air, temperature variation, moisture, and chemical attack.

2.1.2 Steel

Steel, an alloy of Fe and carbon with some other elements depending on the type and quality to be achieved, is one of the most popular packaging options when it comes to strength, ease of fabrication, reusability, and heavy-duty use. Steel is used for the fabrication of strong frames for the packaging and transportation of industrial goods and equipment. Thus, it is also popular for food-contact packaging applications because different grades of steel can offer hygienic and convenient options. Food-containing grades of steel are essentially electrolytic tinplate (ETP) and electrolytic chromium/chromium oxide coated steel (ECCS) (CEN 2001). The Cr coating prevents atmospheric oxidation and sulphur staining and improves lacquer adhesion. ECCS also has an additional organic coating and is normally used for drawn cans, can ends, and lug closures where welding is not required. The Cr coating provides excellent protection against corrosion due to sulfide staining by certain foods. Many times the sulphur present in the food products reacts with electrolytic tinplate as well as with ECCS causing the deposition of black SnS and white FeS, respectively.

2.1.3 Tin and Tinplate

Tin (Sn) is typically used in its pure form or is applied as an additional thin layer on steel used for packaging. Tinplate, on the other hand, is produced from low-carbon steel by coating it with thin layers of Sn. Coating is performed by dipping the steel sheets in molten Sn (hot-dipped tinplate) or by the electro-deposition of Sn on the steel sheet (electrolytic tinplate). Although Sn provides corrosion resistance to steel, tinplate containers are often lacquered to create an inert barrier between the metal and the product, especially if it is used for packing food items. The commonly used lacquers in the process are epoxy phenolic and oleo-resinous groups and vinyl resins. Thus, acting as an excellent barrier against gases, water vapor, light, and odors, tinplate can be heat-treated and sealed hermetically, thus making it effective for packaging sterile products. It is also an excellent substrate for litho-printing and is an outstanding graphical decoration. Its relatively low weight and high mechanical strength make it easy to ship and store. At end use, tinplate can be easily and economically recycled multiple times without loss of quality.

2.1.4 Tin-Free Steel

Tin-free steel often requires a coating of organic material to provide complete corrosion resistance, and it is marginally less expensive than tinplate. Even though the chrome/chrome oxide makes Sn-free steel unsuitable for welding, it offers excellent adhesion of coatings such as paints, lacquers, and inks (Fellows and Axtell 2002).

2.2 Glass

For generations, glass material has been widely used for packaging of reactive substances such as chemicals. Moreover, glass is more a traditional and attractive packaging material when it comes to traditional preparations including drinks and processed food. Glass can be made in different colors and molded into a variety of shapes. Manufacturing and processing of glass involves different metals. Glass is manufactured from sand, soda ash, limestone, and cullet as well as their mixtures. To impart colors in glass, oxides of Fe, chromium (Cr), cobalt (Co), nickel (Ni), and selenium (Se) are used, respectively, for yellow/green, green, blue, violet/brown, and red colours. Lead glazes are widely used on pottery because they are inexpensive and easy to use (Colomban 2005).

2.3 Paper and Paperboards

Paper and paperboard are one of the most common packing materials worldwide (Table 1). Although they are made from wood, plants, and recycled paper and paperboard waste, they contain several metals such as Pb, Cr, Cu, and Ni in very low concentrations as contaminants. Packaging industries mostly use corrugated paperboard for making paper packaging. According to the European Parliament and Council Directive 94/62/EC norm (w.e.f. 2001), the amount of Pb and Cd in packaging materials should not exceed 100 ppm to prevent migration from packaging to food by weight over 5 years (Rozaslin et al. 2010). Metal migration from food packaging and food containers to food and beverages could be due to pH and salt concentrations of the food products and the coating of packaging. Lithopone used in the filled paper coating uses zinc sulfide and barium sulfate. Cd or Zn

Table 1 List of packaging materials from paper and paperboards

Paperboards	Paper
1. Corrugated container 2. Folding cartons 3. Sanitary food containers (milk and beverages, cartons and trays, lipid tight) 4. Fibre and composite packaging (cans, drums) 5. Rigid boxes 6. Moulded pulp products	1. Flexible packing 2. Converted wraps 3. All papers 4. Paper foil 5. Wrappers 6. Specialty bags 7. Label and tags 8. Heavy-duty bags 9. Tapes 10. Wadding

pigments are sometimes used for the improvement of the fluorescence properties of papers, whereas zinc oxide is used to enhance the cohesive strength of paper coatings. During recycling of paper mill sludge, concentrations of metals such as Zn, Pb, Cd, Ni, and Cu increases. Rozaslin et al. (2010) studied heavy metals in raw recycled paper mill sludge and found Cu, Zn, Ni, and Pb concentrations $\leq$88, 251, 26, and 177 mg/kg, respectively.

2.4 Textile

Growing demand for reusable packages and containers is opening new opportunities for textile products in this market. Packaging comprises numerous flexible packaging materials made of textiles are used for packing a variety of commodities for industrial, agricultural, consumer, and other use. Metals, being an essential constituent of most dyes, are used in textiles in low concentrations. For example, C.I. Mordant Black 11, the most common black dye, is Cr based and is used in the coloration of textiles (Rybicki et al. 2004). On-demand customization and supply of sacks and bags made of traditional jute, cotton, or other natural fibers are gradually increasing and are now termed “Packtech”. They can range from heavy weight, densely woven fabrics—such as bags, sacks, flexible intermediate bulk carriers, and wrappings—to light weight, non woven material such as durable papers, tea bags, and wrappings. Other common examples of textile packaging include laundry bags and other bulk-packaging products; sacks for storage; twine and string to tie packages; non-paper tea bags and coffee filters; soaker pads (food); net and woven fiber strapping; lightweight mailbags; and soft luggage.

2.5 Laminates and Metallized Films

Laminates involve the binding of aluminum foil to paper or plastic film, whereas metallized films are plastics containing a thin layer of aluminum for the improvement of barrier properties against moisture, oil, air, and odor (Fellows and Axtell

2002). Lamination is mainly used for high-value foods such as dried soups, herbs, and spices. Although the components of laminates and metallized films are technically recyclable, the difficult processes before recycling include sorting and separating the material, which precludes economically feasible recycling.

2.6 *Metals Composites*

Composites are one of the most recent classes of materials that possess greater strength, are lighter in weight, low in maintenance, and high in durability. Their use is increasing in engineering applications. Composites can be in the form of particulates, fiber-reinforced, or structural. Based on the general composition, composites can be classified into polymer-matrix, metal-matrix, and ceramic-matrix composites. MMCs are important packaging materials being used in the electronics, automobile, and aviation industries. However, their cost-effectiveness remains an area in need of research and development.

Among different types of MMC, "cermet" or "cemented carbides" consisting of ceramic particles in a metal matrix is an important and widely used class of materials. In these composites, tungsten carbide or other similar particles are bound together by high temperature and thus can withstand high temperatures. These can provide protection from high-temperature destruction. Currently, Al and Cu reinforced with high thermal-conductivity carbon fibres and SiC particle-reinforced Al are the most preferred packaging MMC options. Boron fiber-reinforced Al (B/AI) heat sinks are in a few production systems, and particle-reinforced (BeO)/AI has recently been commercialized.

2.7 *Antistatic Packaging*

Antistatic packaging is used to protect the materials from electrostatic charges. It is usually used for the packaging and transportation of sensitive electronic components such as populated printed circuit boards. Antistatic packaging usually comprises plastic polyethylene terephthalate, which is generally metallized, which gives them a characteristic silvery black color (Yam 2009). With the increase in manufacturing and transportation of electronic products and pieces, the use of antistatic packing materials is also increasing, and their after-use management will pose a large challenge in the future.

The metallization of plastic polyethylene terephthalate makes it slightly conductive; hence, the product kept inside such a close packet forms a "Faraday cage," thus preventing it from static charges that otherwise accumulate on other materials being rubbed when bags are handled. The other variants of antistatic packing materials that include different polymers are made of low-charging material that does not allow a build-up of charges, but it cannot protect the packaged item from electrical fields as effectively as the metallized variants do. Table 2

Table 2 Metal-matrix composites for different levels of electronics packaging (based on Zweben 1992)

S. no.	Level	Application	Requirement	Components used for MMC packaging
1	Package	Heat sink/cold pates	Heat dissipation (high thermal conductivity) Low thermal stresses Hermeticity Electromagnetic shielding	Carriers Eectronic packages Microwave packages Photonics packages Laser diode packages
2	Printed circuit boards Package support plate		Heat dissipation Low thermal stresses Vibration (high stiff-ness, damping) Lightweight	Printed circuit boards Printed circuit board heat sinks Package-mounting plates
3	Subsystem (box)		Heat dissipation, insulation vibration and shock (high stiff-ness, strength) Electromagnetic shielding Light eight	Electronic enclo-sures (chassis, black boxes) Covers
4	Support structure		Vibration and shock Lightweight	Support structures

provides the desirable characteristics for the packaging of electronic items and application of MMCs. Disposal of electrostatic packaging materials is commonly through land filling. However, a few variants have entered the market that degrade in approximately 9 months.

3 Post-use Management of Metal-Based Packaging Materials

After use, the packaging material either turns into waste or is reused and recycled depending on its after-use conditions, its properties, and the cost-effectiveness and practical feasibility of recovery and recycling. The most convenient and common method of disposing waste is landfilling or incineration. This technique, however, contaminates the environment by way of leaching to groundwater, surface-water runoff, air emission of toxic contaminants, and deteriorating soil quality. However, when it comes to metals, a major fraction of these are recyclable. Where metals are present in small quantities, such as in textiles and paper, recovery techniques are being developed. In composites, where metals are intricately associated with non biodegradable polymers, methods are being developed to promote their reuse and biodegradation. Most of these are in primary phases of research and development, but they offer potential solutions for the future. In subsequent sections of this chapter, we briefly discuss the implications of metal contamination in the environment resulting from waste disposal as well as developments in the recycling, reuse, and recovery of metal-based packaging.

4 Human Health Effects of Toxic Metals

Metals are worldwide-distributed pollutants and are notable for their tendency to bioaccumulate and biomagnify with their increase in the tropic level. Some metals, such as Cu, Mg, Zn, Fe, etc., are essential for human and plant life and play a significant role in the functioning of enzymatic systems. However, metals exceeding their threshold limit for human consumption or exposure may cause improper functioning of human physiology and metabolism. Other metals, such as Pb, Cd, Hg, As, Cr, etc., have no useful role in human physiology, and moreover they may cause harmful effects on human health. Table 3 shows the effects of various metals on human health.

Table 3 Harmful effects of selected metals on human health

Metals	Adverse effects on human health	References
Iron	Neoplasia, cardiomyopathy, atherosclerosis, and chronic diseases	Gackowski et al. (2002), Kruszewski (2004)
Copper	Weakness, lethargy, anorexia, dysfuction of kidney, liver, and brain; vascular collapse; cirrhosis, obstructive hepatobiliary disease; extrahepatic biliary atresia; neonatal hepatitis, choledochal cysts; and a-1-antitrypsin deficiency	Semple et al. (1960), Winge and Mehra (1990), Beshgetoor and Hambidge (1998)
Zinc	Decreased immunity, lethargy, focal neural deficit, respiratory disorder, nausea/vomiting, epigastric pain, diarrhea, prostate cancer, altered lymphocyte function, Cu deficiency	Mocchegiani et al. (2001), Plum et al. (2010)
Manganese	Psychiatric symptoms such as hyperirritability, violent-acts hallucinations, decrease of libido and incoordination, crippling of extrapyramidal system, neurological disorder, paralytic disease, and pancreatitis	Cotzias et al. (1968)
Magnesium	Muscular weakness, difficulty in breathing, electrocardiogram changes	Workman et al. (2013)
Cobalt	Cardiomyopathy, hypothyroidism, neurological damage, impaired senses, polycythemia, neuropathy, seizures, headaches, liver damage, blindness, cancer	Sauni et al. (2010), Catalani et al. (2012)

(continued)

Table 3 (continued)

Metals	Adverse effects on human health	References
Molybdenum	Increased blood uric acid, gout, pneumoconiosis, liver cirrhosis	Hanaa et al. (2000)
Chromium	Dermatitis, allergy, asthma, eczema, ulcer, gastroenteritis, perforation of the nasal septum, bronchial carcinomas, hepatocellular deficiency, renal oligoanuric deficiency	Baruthio (1992)
Lead	Learning disabilities, behavioral problems, mental retardation, seizures, coma, death at higher doses	Harvey (2002)
Cadmium	Hypertension, arthritis, diabetes, anemia, arteriosclerosis, impaired bone healing, cancer, cardiovascular disease, cirrhosis, reduced fertility, hyperlipidemia, hypoglycemia, headaches, osteoporosis, schizophrenia, strokes	Benoff et al. (2000)
Arsenic	Anorexia, edema, fluid loss, goiter, herpes, interferes with the uptake of folic acid, inhibition of sulfhydryl enzyme systems, jaundice, kidney and liver damage, pallor, peripheral neuritis, stupor, vasodilation, vertigo, vitiligo	Col et al. (1999)
Mercury	Miniamata disease, adrenal gland dysfunction, alopecia, anorexia, ataxia, bipolar disorder, birth defects, dizziness, fatigue, hearing loss, hyperactivity, immune system dysfunction, kidney damage, numbness and tingling, excessive salivation, schizophrenia, thyroid dysfunction, timidity, tremors, peripheral vision loss, blindness	Grandjean et al. (2010), Trasande et al. (2005)
Aluminum	Amyotrophic lateral sclerosis, anemia, colic, fatigue, dental caries, dementia, dialactia, hypoparathyroidism, malfunctioning of the kidney and liver, neuromuscular disorders, osteomalacia, Parkinson's disease, Alzheimar's disease	Yokel (2000), Reddy (2014)
Nickel	Vertigo, cyanosis, tachycardia, palpitations, lassitude, kidney dysfunction, lethargy, ataxia, hypothermia, asthma, bronchitis, rhinitis and pneumoconiosis, and enetic, developmental, immunological, neural, and reproductive defects	Das et al. (2008)

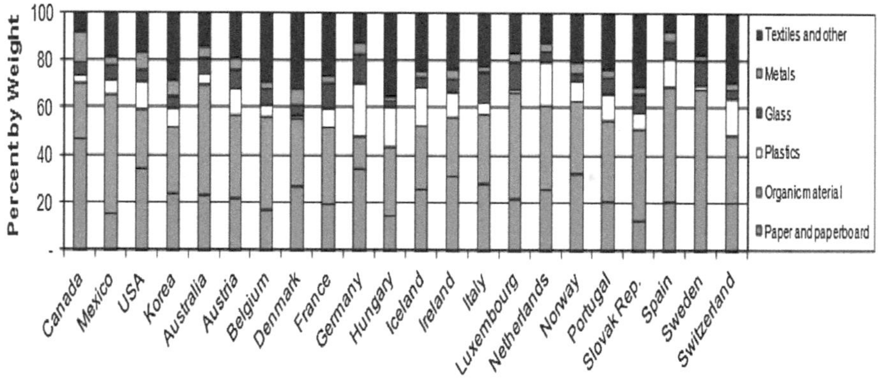

Fig. 1 Municipal solid-waste composition in different countries (adapted from http://faculty.mercer.edu/)

5 Share of Metal-Based Packaging Materials to Waste Composition

There is lack of information on how much metal-containing packaging waste is generated in different parts of the world and from different sectors; however, there are few studies in certain parts of the world focusing mainly on municipal waste. As is clear from Fig. 1, metals are the smallest constituent of waste generated in most countries. Paneque et al. (2008) observed that between 1997 and 2005, packaging waste-generation increased significantly and was directly correlated with an increase in GDP. During this period, metal packaging waste increased by 37.47 %.

6 Management of Metal-Based Packaging Waste

To manage the increasing quantity and complexity of waste generation from different sectors, smart, economic, and environmentally friendly management systems are required. Currently the most common mode of waste management is landfilling incineration. Landfill sites are large areas meant for waste disposal. It is also a cost-effective way of waste management. Contrary to this, resource recovery and incineration both requires extensive investments for infrastructure. Modern landfills are well-engineered facilities with several measures—such as appropriate location, design, operational procedure, and monitoring at regular time intervals—to ensure compliance with federal regulations. Municipal solid-waste landfills receive household waste, packaging waste, non hazardous sludge, industrial solid waste, and construction and demolition debris. In most countries, landfills must comply with certain criteria including the following:

- Establishment of the landfill site in suitable geological areas away from faults, wetlands, flood plains, or other restricted areas;
- Bottom and sides of the landfill must be compacted, e.g., with clay soil, to prevent leaching;
- Treatment and disposal of leachates from landfill;
- Covering of landfill site to reduce odor; control litter, insects, and rodents; and protect public health;
- Regular qualitative analysis of groundwater to monitor leachates (if released);
- Covering of landfill along with long-term management of the landfill site; and
- Control and clean landfill contaminants released and maintain groundwater quality.

Several reports have suggested that metals entering the landfills is large enough to call them "future urban mines" (UNEP-International Resource Panel 2011). For Cu only, the global landfill stockpile is estimated to be 225 million metric tons. Incineration is another common technique of waste disposal. It offers the advantage of volume reduction of wastes and the destruction of much of the organic materials that could contribute to the production of toxic leachates and air emissions during landfilling.

Incineration is a significant technique of volume reduction and concentrates the metals present in waste to be disposed. However, it increases the mobility of the metals present, thus making them more bioavailable to be much more readily absorbed by living organisms, (Denison and Ruston 1990). Generally, incinerated waste is then dumped into landfills.

Although landfills stocks could be potentially reusable, their recovery may not be economically feasible. Several metallurgical techniques have been applied to recover metals from landfills and wastes (Jones et al. 2013). Mostly they consist of pyrometallurgical, hydrometallurgical, and biometallurgical techniques.

(a) Pyrometallurgical processing: Pyrometallurgy is a stepwise process including incineration, smelting, drossing, sintering, melting, and reactions in a gas phase at high temperatures (Gramatyka et al. 2007). The waste material is immersed in a molten metal bath at 1250 °C and churned by 39 % oxygenated air. During processing, plastics and other flammable materials degrade resulting in oxidative conversion of impurities, such as Fe, Pd, and Zn, into oxides fixed in an Si-slag. Thereafter, the metals are recovered from the slag. Cu film containing other precious metals is refined by electrolysis with nearly 99.1 % recovery of Cu along with precious metals such as gold, silver, platinum, palladium, selenium, tellurium, and Ni. However, the integrated smelters used for pyrometallurgy cannot recover Al and Fe because of their negative implications on the properties of the slag. Similarly, ceramics and glasses increase the amount of the slag from blast furnace and reduce the recovery of precious metals. Thus, pyrometallurgy favors the partial recovery of metals from wastes.

(b) Hydrometallurgical processing: During the past few years, hydrometallurgical techniques have been favoured over the pyrometallurgical processing due to its more predictable, easy, controlled processing. The hydrometallurgical processing consists of a series of acidic and alkaline treatments of solid materials. The initial steps involve the extraction of soluble constituents from solid waste using solvents in the forms of cyanide, halide, thiourea, and thiosulfate. Metal recovery from leachates is performed by cementation (Orhan 2005), solvent extraction (Navarro et al. 2007), adsorption onto activated carbon (Alorro et al. 2009), and ion exchange (Vasilyev et al. 2015). Furthermore, the metal-recovery process consists of low-temperature carbonization and roasting of the wastes, leaching with nitric acid solution to remove Ag and other metals, and the use of aqua regia for the extraction of gold.

(c) Biometallurgical processing: This may also be used as an alternative to the above-mentioned processes, although it is still in the very initial stages of application. Biometallurgy is a biotechnological process that utilizes the interaction between microorganisms and metals or metal-bearing substances. For example, solubilizing heterotrophic microorganisms, including bacteria and fungi, which secrete citric, oxalic, and gluconic acids (Henderson and Duff 1963; Avakyan and Robotnova 1971; Valix et al. 2001), can dissociate metallic ores or metals from solid wastes.

Although the extraction of metals from landfill sites may be feasible, recycling is the most preferable and environment friendly option; wherever recycling is not possible, it is recommended that their concentrations in the environment could be maintained below the recommended levels, a challenge or which biorecovery, bioleaching, and biodegradation offer a potential solution. Following sections discuss these waste management techniques.

7 Recycling and Reuse

Pure metallic or alloy wastes are among the most economically recycled wastes, and doing so has been in practice all over the world for a very long time. Graedel et al. (2011) analyzed the information available on metal recycling and found that the most commonly used metals have recycling rates of >50 %. Sorting, cleaning, melting, and casting are the four basic steps in metal recycling. However, the recycling process become complicated and eventually cost-intensive as the complexity of material to be recycled, e.g., composites, increases.

The major challenge in the recycling of the composites is the economic feasibility of the recycling processes and the recovery of various materials. This owes to the complex nature of the composites, differences in the composition of composites used for different products, and the unique physical properties of composites. For example, metals can be reclaimed from different alloys by subjecting wastes to high temperatures, which melt and separate into different components. This technique, however, does not work on thermoset composites.

Although technological advancements have made composite recycling more practical, lowering the price of the process and creating a suitable market for these recycled and reclaimed components remain challenging. However, a few companies have emerged that are making profits with the chopped and milled carbon fibres recovered from the aerospace industry.

Recycling technologies have also been developed for other types of composite materials such as thermoplastic matrix- and metal matrix-based composites. The common methods of recycling and metal recovery techniques from these materials include remelting–casting for MMC and direct remelting for dirty scrap. Remelting-casting of MMC is relatively costlier than the same technique being followed for alloys or reinforcements. For foundary scrap, direct remelting is performed under the inert atmosphere of dry Ar. To reuse MMC and dirty scrap, a combination of remelting, fluxing, and degassing cleaning are performed. Very dirty scrap, on the other hand, is used only for recovering metals following the techniques of remelting and refining to separate reinforcement from Al alloys. During various processes, ferrous metals are removed through magnetic separation. The techniques vary according to the MMC properties and compositions.

7.1 Recycling of Metal-Matrix Composites (MMC)

MMC materials have relatively much higher economic values than its constituent base metals or alloys. This makes the recycling of the MMCs economically infeasible, but their direct reuse is preferred. When the MMC packages become dirty or old after a single or several uses, their qualities can be restored to a degree by fluxing and cleaning by degassing. However reuse cannot go on indefinitely, and hence better recycling solutions become an important area of research and development.

Recovering metals from MMCs makes use of mechanical and chemical methods. In the mechanical method, the matrix metal is squeezed or filtered out of the composite after melting. The chemical method employs the use of a molten flux to absorb and wet reinforcement particles to facilitate easy separation from the molten metal. Electrorefining is carried out in ionic liquid composed of 1-butyl-3-methylimidazolium chloride and anhydrous $AlCl_3$. For some Al-based MMCs, Al metal or alloys are recovered through remelting. In the case where recycling of the whole material is not practical, the matrix metal is recovered by melting.

Another recycling approach is to use alternating layers of metal foils and fibre/matrix resin stacked and then consolidated under pressure, which leads to the formation of fibre–metal laminates usable in certain industries, most notably the automobile and aviation industries. The first such laminates, called ARALL (aramid-reinforced Al laminate), was produced by Vogelsang et al. (1981) and consisted of Al sheets and aramid fibre/epoxy prepreg. These laminates are now known as GLAss Reinforced (GLARE) FML (Vlot 2001). For example, GLARE FML as already been used in an Airbus model.

When it comes to recycling FMLs, their low production cost makes them a poor candidate for recycling. Although research is directed toward its recycling and reuse, due to the low market value of epoxy resin and glass fibre, only Al recovery is the main focus in the recycling of GLARE FML (Tempelman 1999). For this, GLARE is delaminated by cryogenic liberation to separate the Al foils from the epoxy resin and glass fibres. The next step is to subject the mixture of liberated Al and unseparated GLARE FML to an eddy current separator. The cost of low-temperature cryogenic liberation is high compared with the market value of the recovered Al scrap. For this, delamination occurs at high temperature of approximately 220 to 500 °C in an open furnace, which destroys the epoxy resin. Delamination in a fluidised bed reactor is another possibility. After thermal delamination, relatively clean glass fibres and Al plates are generated. For effective separation of the matrix metal/alloy from the reinforcement fibres and filaments, a mixture of NaCl and KCl—along with some fluorides such as Na_2SiF_6 and NaF—are used in molten form as a flux (Nishida 2001). Melting is conducted inside furnaces of different shapes and based on different techniques. The most common types are induction furnaces, reverberatory melters, hearth furnaces, and rotary barrel furnaces. The recycling and recovery of usable portions of metal and nonmetal components of carbon fiber are also being attempted. Boeing Company presents a great example: It recovers and recycles the scraps of its retired planes.

For thermosetting composites, reclamation is a three-step process comprised of the first thermal pretreatment followed by two wet chemical processes. Tertiary recycling reclaims fibres, thermoplastics, and thermosetting polymers. The thermosetting polymers are recovered into usable hydrocarbon fractions, which serve as building materials for new polymers, fuels, and chemicals. Various methods are employed to reduce size (crushing, chopping, drying, etc.), perform off-gas treatment and distillation, and recover metals, fibers, and carbon chars.

The most promising way of utilizing thermoset composites is to use them as filler materials in combination with conventional fillers, such as asphalt, after grinding them into granules. On the other hand, reclaimed short fibres are used to reinforce sheet molding compound and bulk molding compounds.

To promote the recycling and reuse of composites, most of the European Union (EU) member countries forbade landfill disposal of composites in 2004. In the United Kingdom, the planning of recyclability of components after their end use must be considered at the time of designing the product.

8 Toward Biodegradable Packaging Options of Metal Composites

One such composite is to use cellulose as the matrix. Besides being the most abundant and widespread biopolymer on the earth, cellulose also holds specific properties that make it a suitable candidate for the development of environmentally friendly, biocompatible, and functional composites. The molecular structure of cellulose and its tendency to form intramolecular and intermolecular bonding give it suitable properties to be used into composites. To make cellulose-based MMCs, varieties of metal nanoparticles can be used as dispersed phases in cellulosic bio-nanocomposites (Dankovich and Gray 2011; Wu and Fang 2003; Ma and Fang 2006). The synthesis of colloidal metal nanoparticles has received great attention, and significant advancement in their synthesis has been made during the last decade. Their large specific surface area and unique optical, electronic, magnetic, and antimicrobial properties have introduced them as a potential option in composite materials. Silver is one of the most common metals used in these biocomposites.

With proven efficacy as antimicrobial materials, silver nanoparticles have led researchers to formulate silver nanoparticles-cellulose matrix composites (Pinto et al. 2009), which can be used for antibacterial medical- and food-packaging materials.

9 Bioleaching

Bioleaching is the process to help solubilizing the metals from solid phase using biological processes such as the use of acid-producing microorganisms, fungi (*Astraeus odoratus*; Kumla et al. 2014), and sulfur-oxidizing bacteria (*Acidithiobacillus thiooxidans* and *Thiobacillus oxidans*; Dhakar et al. 2015). It is a cost-effective way of recovering metals from a much diffused state such as when they are present in compost or waste mixes, landfill, and soil (Pathak et al. 2009). Similarly, iron-oxidizing bacteria such as *T. ferooxidans* and *Leptospirillum ferooxidans* are used to oxidize ferrous compounds (Johnson and Mc Ginness 1991). Bioleaching techniques can be used in both direct and indirect ways. Direct mechanisms are employed generally to leach metal sulfide. Metals in wastes often get transformed into respective sulfides under reducing conditions of the waste layer. Tateda et al. (1998) reported that 0.7 % metals such as Cd, Cu, Pb, and Zn can be easily recovered by >50 % under the action of sulfur-oxidizing bacterium *Thiobacillus thiooxidans*. Bioleaching processes also depend on the physicochemical properties of the material, which affect metal solubility and microbial activities such as pH. Krebs et al. (2001) found that eight semicontinuous inoculations of the waste ash with *T. thiooxidans* resulted in high leach ability, e.g., >80 % for Cd, Cu, and Zn; 60 % for A; and 30 % for Fe and Ni.

10 Biosorption

Biosorption refers to the passive (i.e., not metabolically mediated) uptake of metal or nonmetal species by living or dead biomass (Fig. 2). This process plays a significant role in the removal and recovery of metals from wastes. Compared with other techniques, such as precipitation and synthetic ion exchange resins,

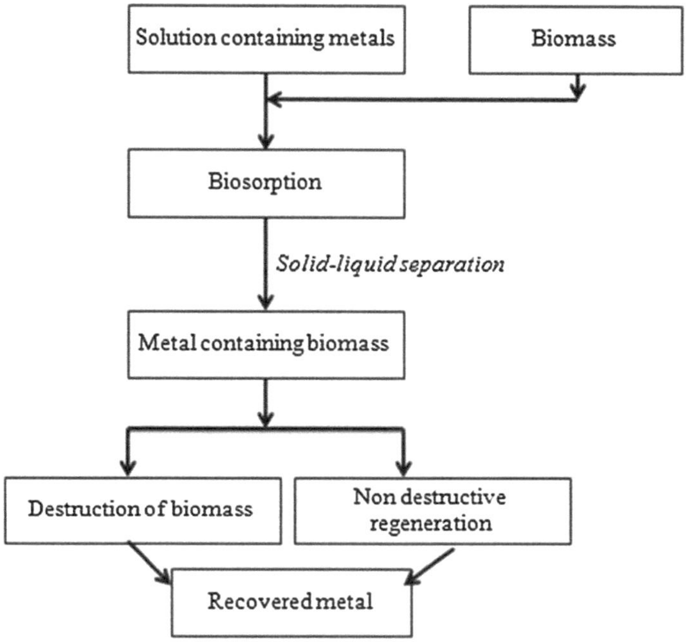

Fig. 2 Biosorption of metals from wastes using biosorbent material (based on Araujo et al. 2013)

biosorption is advantageous as a cost-effective and more efficient option (Volesky and Naja 2005). It encompasses a physicochemical mechanistic approach where metal species are removed from an aqueous medium by microbial biomass and certain products (Fomina and Gadd 2014).

A variety of microbial organisms and other biomass options have been reported to have good biosorption potential in removing metals such as Pb and Cr from wastes (Orhan and Buyukgungor 1993; Bahadir et al. 2007). For example, several species of algae (red and brown algae) possess high metal-binding capacity (Schiewer and Volesky 2000) whereby their cell walls bind to metals. The presence of carboxyl and sulphate groups in algal cell wall acts as active sites for metal binding. Alginate and fucoidan in brown algae are known for their metal-binding properties (Davis et al. 2003). The biosorption properties greatly depend on the environment, and pH plays a main role. Both carboxyl and sulphate groups become protonated at low pH and therefore become less available for binding metals. In addition to pH value, ionic strength also plays an important role in the process. Wastes are generally characterized by greater concentrations of sodium, thereby increasing the ionic strength and hence reducing biosorption (Greene et al. 1987; Ramelow et al. 1992) of weakly bound metals (Zn, Ni). However, strongly bound metals are unaffected by greater Na concentrations. Among diverse algal species, *Petalonia* sp. and *Sargassum* sp. appear to be the most promising biosorbing agents due to the presence of higher metal-binding sites (Schiewer and Wong 2000).

11 Metal Degradation

Metals corrode naturally through an electrochemical process, but this process can be enhanced by certain biological activities (Breslin 1993; Gu et al. 1998a, b; Ford and Mitchell 1990). As shown in Fig. 3, the microbial habitat on the surface of a metal forms a differential aeration zone under aerobic condition, which results in an electrochemical gradient at the interface of the microbial biofilm and the surface of the metal. The area exposed to oxygen serves as a cathode, whereas the area beneath the biofilm serves as anode. Electrons are transported from the anode to the cathode due to the electrochemical gradient resulting in metal dissolution, crevice corrosion, and pitting (Gu et al. 2000; Videla 1996; Ford and Mitchell 1990). Furthermore, the decline in the oxygen level results in the establishment of an anaerobic zone, thus supporting the growth of anaerobic microbes. These anaerobic microbial communities cause corrosion of underlying metals by cathode depolarization. Under anaerobic conditions, methanogenic microorganisms also participate in metal corrosion (Daniels et al. 1987). This biological process can be used for degrading metals present in the environment. A wide range of aerobic and anaerobic microorganisms cause biodegradation of metal alloys by the process of corrosion. Amongst aerobic microorganisms, sulphate-reducing (SRB), thermophilic, Fe-oxidizing, exopolymer, and acid-producing bacteria are the most common metal degraders. The mechanism involved is either metal transformation or complex formation, including the functional groups of exopolymers, with the release of metal species in the solution (Chen et al. 2006; Gu et al. 2000; Dexter 1993; Ford and Mitchell 1990).

In presence of the Fe-oxidizing bacteria (*Sphaerotilus* sp., *Leptothrix* sp., *Gallionella* sp., and *Siderocapsa* sp.) under oxygenic conditions and neutral pH, Fe^{2+} is oxidized to Fe^{3} thereby increasing the rate of Fe degradation. Fe_3O_4 thus formed is deposited enzymatically by *Gallionella ferruginea* and nonenzymatically by *Leptothrix* sp., *Siderocapsa* sp., and *Siderococcus* sp. (Ehrlich 1996). Similarly, Mn deposition takes place by the action of bacteria such as *Aeromonas* sp., *Caulobacter* sp., *Caulococcus* sp., *Citrobacter* sp., *Clonothrix* sp.,

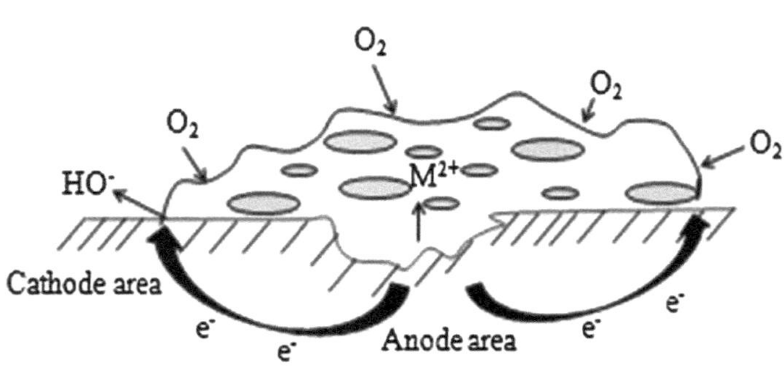

Fig. 3 Release of metal (M^{2+}) ions from anode area due to corrosion by microbial biofilm

Flavobacterium sp., *Pseudomona* sp., *Streptomyces* sp., and *Vibrio* sp. (Dickinson et al. 1996; Olesen et al. 1998).

Nonferrous metals are acted on by sulphate-reducing bacteria, which immobilize and precipitate them (Sakaguchi et al. 1993). This technique is particularly helpful in the conversion of toxic to nontoxic forms of metals. For example, the more toxic form Cr^{6+} is oxidized to the less toxic metal Cr^{3+} under the action of a range of microbes including *Aeromonas dechromatica, Agrobacterium radiobacter, Arthobacter* sp., *Bacillus subtilis, B. cereus, Desulfovibrio vulgaris, Eschericia coli, Enterobacter cloacae*, and *Flavobacterium devorans* (Ehrlich 1996).

12 Phytoremediation of Heavy Metals from Landfill Sites

Vegetation cover at landfill sites is effective in controlling erosion and the removal of contaminants and in the treatment of leachates (Maurice 1998). Phytoremediation is a plant-based technology that uses plants for the removal of pollutants such as metals, pesticides, solvents, explosives, crude oil and its derivatives, and various other contaminants from environmental media (air, soil and water) (Mc Cutcheon and Schnoor 2003). It is cost-effective facilitating easy monitoring of plant performance and potentially a least harmful method of metal removal that preserves the environment in its natural state. Phytoextraction, phytostabilisation, phytotransformation, phytostimulation, phytovolatilization, and rhizofiltration are other phytoremediation techniques. A generalized picture of the concept is given in Fig. 4.

12.1 Phytoextraction

Phytoextraction is the process of the removal of contaminants from contaminated soil, sediments, or water and their storage in harvestable plant biomass. Water hyacinth (*Eichhornia crassipes*) grown in tap water supplemented with 0.35, 0.70, and 1.05 mg l^{-1} of Cu or 0.27, 0.54, and 0.81 mg l^{-1} of Cd for 25 days effectively extracted approximately >90 % of Cu and Cd (Swain et al. 2014).

12.2 Phytostabilisation

This reduces the mobility of metals, thereby stabilizing them in the substrate or roots. For instance, poplar tree plantation is an effective tool in immobilizing water-soluble contaminants and arresting heavy metals at a contaminated site (Schnoor 2000).

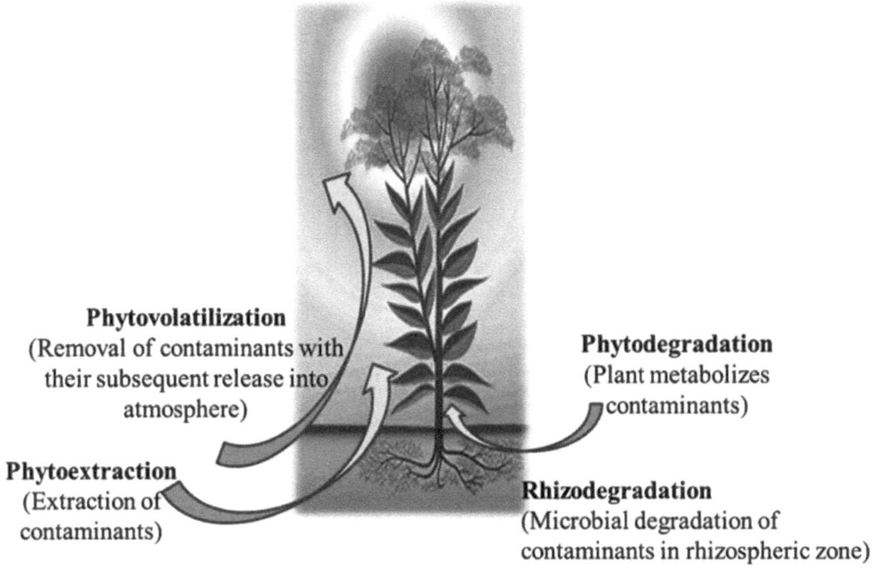

Fig. 4 Phytoremediation techniques for the removal of heavy metals from contaminated sites

12.3 Phytostimulation

An enhancement of soil microbial activities for the degradation of contaminants, typically in association with a rhizospheric zone, is termed "phytostimulation". Bacterial associations in the rhizospheric zone, which usually are also considered as plant growth-promoting bacteria, decrease metal toxicity to plants. Common bacterial species identified as phytostimulants are *Bacillus* sp., *Pseudomonas* sp., *Azotobacter* sp., *Rhizobium* sp., *Klebsiella* sp., and *Paenibacillus polymyxa* (Joseph et al. 2007; Phi et al. 2010).

Phosphate-solubilizing bacteria—such as *Rhodococcus* sp., *Arthrobacter* sp., *Serratia* sp., *Chryseobacterium* sp., *Gordonia* sp., *Phyllobacterium* sp., *Delftia* sp., *Xanthomonas* sp., *Azotobacter* sp., *Klebsiella* sp., *Vibrio proteolyticus*, *Enterobacter* sp., and *Pantoea* sp.—have also been found to be very effective in reducing metal toxicities in plants (Wani et al. 2005; Chen et al. 2006; Kumar et al. 2001; Chung et al. 2005; Vazquez et al. 2000).

12.4 Phytovolatilization

The process of uptake and release of the contaminant or a modified form of the contaminant into the atmosphere by transpiration is known as "phytovolatilization". It is not a very significant removal process for most metals; only for selected metals, such as Hg and Se, has it been reported as effective. For the uptake of Se by plants and its bioconversion into nontoxic gas, dimethyl selenide has been reported as effective (Terry et al. 1995).

12.5 Rhizofiltration

Rhizofiltration is the process where roots or whole plants absorb metals from polluted effluents and are later harvested to diminish metals in the effluents. Hence, it is primarily used to remediate extracted groundwater, surface water, and wastewater with low levels of metals (Ensley 2000).

12.6 Remediation and Biodegradation Potential of Earthworm Species

Leachates from landfill sites are the prime source of toxic and persistent metals. Remobilization of these toxic metals may harm both humans and our ecosystem. There are some bioprocessing methods to manage solid wastes. Vermicomposting of wastes has been recognized as a preferential option to stabilize various kinds of wastes (Negi and Suthar 2013). Gut microflora of earthworm degrades waste materials to finer substrates by mineralizing the organically bound nutrients into bioavailable form with the release of mineral nutrients through excreta. In addition, earthworm remediates heavy metals in processed products by accumulating metals in their intestine as metal-bound protein metallothionein (Sahariah et al. 2015). After the death of earthworm, metal-bound protein molecules exposed to the soil environment are retained in humic substances in immobilized forms (Nannoni et al. 2011). Saharia et al. (2015) reported a significant reduction in heavy-metal content after vermicomposting of MSW with cow-dung amendment.

13 Conclusion

Metals have been a preferred choice for packaging and storage since ancient times. Metals such as Fe, Al, Sn, and Cu were used as boxes, cans, and cylinders for containing and facilitating the transport of goods. In modern times, other metals, such as Pb, Cd, and Cr, in pure form or as alloys have been used, and are also mixed with other materials such as polymers, fibres, and plastics to produce next-generation packaging materials such as composites and laminates. These different metal-based materials enable the protection, transportation, and storage of variety of goods ranging from food material (e.g., cans, secure caps, food processing, and microbially protected containers) to sensitive electronic components (e.g., electrostatic packaging materials). However, with our increasing usage of packaging material, the generation of associated waste is also increasing. Common disposal techniques of packaging-related wastes are dumping them into landfills and incineration. Because metals can contaminate the environment and lead to serious human health hazards, metal-elated wastes are recycled, reused, and recovered. Due to their long life and retention of original material properties even after use, pure metals and alloys are easily recycled by remelting, purification, and casting. Composites and laminates are not economically beneficial to recycle, but methods have been developed to recover metals and other useful components based on different physical, chemical, and biological separation techniques at different stages of waste management. Among the different approaches, biotechnological methods offer technically as well as economically potential ways to recover metals from packaging wastes, landfills, and sites contaminated by metals. Among these, the development of biodegradable composites and biosorption appears to be a promising techniques that can improve the sustainable use of metal-based packaging by different sectors.

Acknowledgement Divya Pandey and Meenu Gautam are thankful to the Council of Scientific and Industrial Research, India, for research fellowships in the form of Research Associateship and Junior Research Fellowship, respectively.

References

Alorro RD, Hiroyoshi N, Ito M, Tsunekawa M (2009) Recovery of heavy metals from MSW molten fly ash by CIP method. Hydrometallurgy 97(1):8–14

Araujo CS, Carvalho DC, Rezende HC, Almeida IL, Coelho LM, Coelho NM, Alves VN (2013) Bioremediation of waters contaminated with heavy metals using moringaoleifera seeds as biosorbent. In: Patil YB, Rao (eds) Applied bioremediation—active and passive approaches, p 225

Avakyan ZA, Robotnova IL (1971) Comparative toxicity of free ions and complexes of copper with organic acids for *Candida ulitis*. Microbiology 40:262–266

Bahadir T, Bakan G, Altas L, Buyukgungor H (2007) The investigation of lead removal by biosorption: an application at storage battery industry wastewaters. Enzyme Microb Tech 41(1):98–102

Baruthio F (1992) Toxic effects of chromium and its compounds. Biol Trace Elem Res 32(1–3):145–153

Benoff S, Jacob A, Hurley IR (2000) Male infertility and environmental exposure to lead and cadmium. Hum Reprod Update 6(2):107–121

Beshgetoor D, Hambidge M (1998) Clinical conditions altering copper metabolism in humans. Am J Clin Nutr 67:1017S/1021S

Breslin VT (1993) Degradation of starch-plastic composites in a municipal solid waste landfill. J Environ Polym Degr 1(2):127–141

Catalani S, Rizzetti MC, Padovani A, Apostoli P (2012) Neurotoxicity of cobalt. Hum Exp Toxicol 31(5):421–437

Harvey B (2002) Managing elevated blood lead levels among young children: Recommendations from the advisory committee on childhood lead poisoning prevention. Centers for Disease Control and Prevention. United States Department of Health and Human Services, Atlanta, GA, pp 128

Chen YP, Rekha PD, Arun AB, Shen FT, Lai WA, Young CC (2006) Phosphate solubilizing bacteria from subtropical soil and their tricalcium phosphate solubilizing abilities. Appl Soil Ecol 34:33–41
Chung H, Park M, Madhaiyan M, Seshadri S, Song J, Cho H, Sa T (2005) Isolation and characterization of phosphate solubilizing bacteria from the rhizosphere of crop plants of Korea. Soil Biol Biochem 37:1970–1974
Col M, Col C, Soran A, Sayli BS, Ozturk S (1999) Arsenic-related Bowen's disease, palmar keratosis and skin cancer. Environ HlthPerspect 107(8):687–689
Colomban P (2005) "Glasses, glazes and ceramics—recognition of the ancient technology", Raman spectroscopy. In: Edwards HGM, Chalmers JM (eds) Archaeology and art history. Royal Society of Chemistry, Cambridge, pp 192–206
Cotzias GC, Horiuchi K, Fuenzalida S, Mena I (1968) Chronic manganese poisoning: clearance of tissue manganese concentrations with persistence of the neurological picture. Neurology 18:376–382
Cutter CN (2002) Microbial control by packaging: a review. Crit Rev Food Sci 42(2):151–161
Daniels L, Belay N, Rajagopal BS, Weimer PJ (1987) Bacterial methanogenesis and growth from CO_2 with elemental iron as the sole source of electrons. Science 237(4814):509–511
Dankovich TA, Gray DG (2011) Bactericidal paper impregnated with silver nanoparticles for point-of-use water treatment. Environ Sci Technol 45(5):1992–1998
Das KK, Das SN, Dhundasi SA (2008) Nickel, its adverse health effects and oxidative stress. Indian J Med Res 128(4):412
Davis TA, Volesky B, Mucci A (2003) A review of the biochemistry of heavy metal biosorption by brown algae. Water Res 37:4311–4330
Denison R, Ruston J (1990) Recycling and incineration: evaluating the choices. Island Press
Dexter SC (1993) Role of microfouling organisms in marine corrosion. Biofouling 7(2):97–127
Dhakar S, Ali N, Chauhan RS, Yadav V (2015) Potentiality of acidithiobacillus thiooxidans in microbial solubilization of phosphate mine tailings. J Degraded Mining Lands Manage 2(3):355–360
Dickinson WH, Caccavo F, Lewandowski Z (1996) The ennoblement of stainless steel by manganic oxide biofouling. Corros Sci 38(8):1407–1422
Ehrlich HL (1996) Geomicrobial processes: a physiological and biochemical overview. Geomicrobiology. Marcel Dekker, New York, pp 108–142
Elinder CG, Sjögren B (1986) Aluminium. In: Friberg L, Nordberg GF, Vouk VB (eds) Handbook on the toxicology of metals, 2nd edn. Elsevier, Amsterdam, pp 1–25
Ensley BD (2000) Rationale for use of phytoremediation. In: Phytoremediation of toxic metals: using plants to clean up the environment. Wiley, New York
European Committee for Standardisation (CEN) (2001) European standard EN 10202:2001 cold reduced tin mill products–electrolytic tinplate and electrolytic chromium/chromium oxide coated steel. CEN, Brussels
European Committee for Standardisation (CEN) (2004) EN 602:2004 aluminium and aluminium alloys—Wrought products—chemical composition of semi products used for the fabrication of articles for use in contact with foodstuff. CEN, Brussels
Fellows P, Axtell B (2002) Packaging materials. In: Fellows P, Axtell B (eds) Appropriate food packaging: materials and methods for small businesses. ITDG Publishing, Essex, pp 25–77
Fomina M, Gadd GM (2014) Biosorption: current perspectives on concept, definition and application. Bioresource Technol 160:3–14
Ford T, Mitchell R (1990) The ecology of microbial corrosion. Springer, New York, pp 231–262
Gackowski D, Kruszewski M, Banaszkiewicz Z, Jawien A, Olinski R (2002) Lymphocyte labile iron pool, plasma iron, transferrin saturation and ferritin levels in colon cancer patients. Acta Biochim Pol 49:269–272
Graedel TE, Allwood J, Birat JP, Buchert M, Hagelűken C, Reck BK, Sibley SF, Sonnemann G (2011) What do we know about metal recycling rates? USGS Staff—Published Research. Paper 596. http://digitalcommons.unl.edu/usgsstaffpub/596

Gramatyka P, Nowosielski R, Sakiewicz P (2007) Recycling of waste electrical and electronic equipment. J Achievements Mater Manuf Eng 20(1–2):535–538
Grandjean P, Satoh H, Murata K, Eto K (2010) Adverse effects of methylmercury: environmental health research implications. Environ Health Perspect 118(8):1137–1145
Greene B, McPherson R, Darnall D (1987) Algal sorbents for selective metal ion recovery. In: Metals Speciation, Separatio Fomich J E (2011). A History of Packaging CDFS-133, Ohio State University Fact Sheet, Community Development, Columbus, Ohio. http://ohioline.osu.edu/cdfact/0133.html
Gu JD, Ford TE, Mitton DB, Mitchell R (2000) Microbial corrosion of metals. The Uhlig Corrosion Handbook
Gu JD, Mitton DB, Ford TE, Mitchell R (1998a) Microbial degradation of polymeric coatings measured by electrochemical impedance spectroscopy. Biodegradation 9(1):39–45
Gu JD, Roman M, Esselman T, Mitchell R (1998b) The role of microbial biofilms in deterioration of space station candidate materials. Int Biodeter Biodegr 41(1):25–33
Hanaa M, Eweida A, Farag A (2000) Heavy metals in drinking water and their environmental impact on human health. Paper presented at international conference on environmental hazards mitigation, Cairo University, Egypt, pp 542–556
Henderson ME, Duff RB (1963) The release of metallic and silicate ions from minerals, rocks, and soils by fungal activity. J Soil Sci 14(2):236–246
Holdsworth SD, Simpson R (2007) Thermal processing of packaged foods. Springer Science & Business Media
Hook P, Heimlich JE (2011) A history of packaging CDFS-133. Ohio State University Columbus, Ohio. http://ohioline.osu.edu/cdfact/0133.html. Accessed 28 May 2015
Johnson DB, Mc Ginness S (1991) A highly efficient and universal solid medium for growing mesophilic and moderately thermophilic, iron-oxidizing, acidophilic bacteria. J Microbiol Meth 13(2):113–122
Jones PT, Geysen D, Tielemans Y, Van Passel S, Pontikes Y, Blanpain B, Hoekstra N (2013) Enhanced Landfill Mining in view of multiple resource recovery: a critical review. J Cleaner Production 55:45–55
Joseph B, Patra RR, Lawrence R (2007) Characterization of plant growth promoting rhizobacteria associated with chickpea (Cicerarietinum L.). Int J Plant Prod 2:141–152
Krebs W, Bachofen R, Brandl H (2001) Growth stimulation of sulfur oxidizing bacteria for optimization of metal leaching efficiency of fly ash from municipal solid waste incineration. Hydrometallurgy 59(2):283–290
Kruszewski M (2004) The role of labile iron pool in cardiovascular diseases. Acta Biochim Pol 51:471–480
Kumar V, Behl RK, Narula N (2001) Establishment of phosphate-solubilizing strains of Azotobacter chroococcum in the rhizosphere and their effect on wheat cultivars under greenhouse conditions. Microbiol Res 156:87–93
Kumla J, Suwannarach N, Bussaban B, Matsui K, Lumyong S (2014) Indole-3-acetic acid production, solubilization of insoluble metal minerals and metal tolerance of some sclerodermatoid fungi collected from northern Thailand. Ann Microbiol 64(2):707–720
Ma W, Fang Y (2006) Experimental (SERS) and theoretical (DFT) studies on the adsorption of p-, m-, and o-nitroaniline on gold nanoparticles. J Colloid Interface Sci 303(1):1–8
Maurice C (1998) Landfill Gas Emission and Landfill Vegetation. Lulea University of Technology, Lulea, Sweden
McCutcheon SC, Schnoor JL (2003) Overview of phytotransformation and control of wastes. Transformation and control of contaminants, Phytoremediation, p 987
Mocchegiani E, Giacconi R, Cipriano C, Muzzioli M, Fattoretti P, Bertoni-Freddari C, Isani G, Zambenedetti P, Zatta P (2001) Zinc-bound metallothioneins as potential biological markers of ageing. Brain Res Bull 55:147–153

Nannoni F, Protano G, Riccobono F (2011) Uptake and bioaccumulation of heavy elements by two earthworm species from a smelter contaminated area in northern Kosovo. Soil Biol Biochem 43:2359–2367

Navarro R, Guzman J, Saucedo I, Revilla J, Guibal E (2007) Vanadium recovery from oil fly ash by leaching, precipitation and solvent extraction processes. Waste Manage 27(3):425–438

Negi R, Suthar S (2013) Vermistabilization of paper mill wastewater sludge using *Eiseniafetida*. Bioresource Technol 128:193–198

Nishida Y (2001) Recycling of metal matrix composites. Adv Eng Mater 5(3):315–317

Olesen BH, Avci R, Lewandowski Z (1998) Ennoblement of stainless steel studied by X-ray photoelectron spectroscopy (No. CONF- 980316). NACE International, Houston, USA

Orhan G (2005) Leaching and cementation of heavy metals from electric arc furnace dust in alkaline medium. Hydrometallurgy 78(3):236–245

Orhan Y, Buyukgungor H (1993) The removal of heavy metals by using agricultural wastes. Water Sci Technol 28:247–255

Paneque A, Cháfer C, Pacheco B, Hortal M, Capuz S (2008) Análisis de la situación de la genercíón y gestión de residuos de envaseenEspaña. In: 12th international conference on project engineering, Zaragoza, Spain

Pathak A, Dastidar MG, Sreekrishnan TR (2009) Bioleaching of heavy metals from sewage sludge: a review. J Environ Manage 90(8):2343–2353

Phi Q-T, Park Y-M, Seul K-J, Ryu C-M, Park S-H, Kim S-H, Ghim S-Y (2010) Assessment of root-associated *Paenibacillus polymyxa* groups on growth promotion and induced systemic resistance in pepper. J Microbiol Biotechnol 20:1605–1613

Pinto RJB, Marques PAAP, PascoalNeto C, Trindade T, Daina S, Sadocco P (2009) Antibacterial activity of nanocomposites of silver and bacterial or vegetable cellulosic fibers. Acta Biomater 5(6):2279–2289

Plum LM, Rink L, Haase H (2010) The essential toxin: impact of zinc on human health. Int J Environ R Public Heal 7(4):1342–1365

Ramelow GJ, Fralick D, Zhao YF (1992) Factors affecting the uptake of aqueous metalions by dried seaweed biomass. Microbios 72(291):81–93

Reddy MA (2014) Toxic metals and environmental pollution. In: national seminar on impact of toxic metals, minerals and solvents leading to environmental pollution-2014. J Chem Pharmaceut Sci. 3:175–181. (Special issue)

Rozaslin A, Fauziah CI, Rasidah Wan W, Rosenani AB (2010) Presented paper at 19th world congress of soil science, solutions for a changing world, Brisbane, Australia, 1–6 Aug 2010. http://www.ilsi.org/Europe/Documents/MERTOGLUELMAS.pdf. Accessed on 25 April 2015

Rybicki E, Swiech T, Lesniewska E, Albinska J, Szynkowska MI, Paryjczak T, Sypniewski S (2004) Changes in hazardous substances in cotton after mechanical and chemical treatments of textiles. Fibres Text East Eur 12:67–73

Sahariah B, Goswami L, Kim KH, Bhattacharyya P, Bhattacharya SS (2015) Metal remediation and biodegradation potential of earthworm species on municipal solid waste: a parallel analysis between *Metaphireposthuma* and *Eiseniafetida*. Bioresource Technol 180:230–236

Sakaguchi T, Burgess JG, Matsunaga T (1993) Magnetite formation by a sulphate-reducing bacterium. Nature 365:47–49

Sauni R, Linna A, Oksa P, Nordman H, Tuppurainen M, Uitti J (2010) Cobalt asthma—a case series from a cobalt plant. Occup Med 60(4):301–306

Schiewer S, Volesky B (2000) Environmental microbe—metal interactions. In: Lovley D (ed) ASM Press, Washington, DC

Schiewer S, Wong MH (2000) Ionic strength effects in biosorption of metals by marine algae. Chemosphere 41(1):271–282

Schnoor JL (2000) Phytostabilization of metals using hybrid poplar trees. In: Raskin I, Ensley BD (eds) Phytoremediation of toxic metals: using plants to clean up the environment. Wiley, New York, pp 133–150

Semple AB, Parry WH, Philips DE (1960) Acute copper poisoning: an outbreak traced to contaminated water from a corroded geyser. Lancet 2:700/701
Swain G, Adhikari S, Mohanty P (2014) Phytoremediation of copper and cadmium from water using water hyacinth, *Eichhornia crassipes*. Int J Agric Sci Technol 2:1–7. www.seipub.org/ijast, doi:10.14355/ijast.2014.0301.01
Tateda M, Ike M, Fujita M (1998) Heavy metal extraction from municipal solid waste incineration fly ash using a sulfur oxidizing bacterium. Resour Environ Biotechnol 2(2):137–151
Tempelman E (1999) Sustainable transport and advanced materials. Ph.D. thesis, Delft University of Technology, March 1999, Netherlands
Terry N, Zayed A, Pilon-Smits E (1995) Bioremediation of selenium by plant volatilization (No. CONF-9509139–). In: Abstract of the 7th on ACS special symposium: emerging technologies in hazardous waste management, Atlanta, GA (United States), 17–20 Sep 1995
Trasande L, Landrigan PJ, Schechter C (2005) Public health and economic consequences of methyl mercury toxicity to the developing brain. Environ Health Persp 590–596
UNEP-International Resource Panel (2011) Metal stocks in society: scientific synthesis recycling rates of metals: a status report UNEP, Nairobi, Kenya. http://www.unep.org/pdf/Metals_Recycling_Rates_Summary.pdf. Accessed 28 May 2015
US-Environmental Protection Agency (2010) Municipal solid waste generation, recycling and disposal in the United States: facts and figures for 2010. www.epa.gov/osw/nonhaz/municipal/pubs/2012_msw_fs.pdf. Accessed 17 June 2015
Valix M, Usai F, Malik R (2001) Fungal bio-leaching of low grade laterite ores. Miner Eng 14(2):197–203
Vasilyev F, Virolainen S, Sainio T (2015) Synthesis of hydrometallurgical processes for valorization of secondary raw materials using ant colony optimization and key performance indicators. Hydrometallurgy 153:121–133
Vazquez P, Holguin G, Puente M, Lopez-cortes A, Bashan Y (2000) Phosphate solubilizing microorganisms associated with the rhizosphere of mangroves in a semi-arid coastal lagoon. Biol Fertil Soils 30:460–468
Videla HA (1996) Corrosion inhibition by bacteria. In: McCombs K, Albert W, Starkweather Jr (eds) Manual of biocorrosion, USA, pp 121–135
Vlot (2001) Historical overview. In: Vlot A, Gunnink JW (eds) Fibre metal laminates, an introduction. Kluwer Academic Publishers, London, pp 3–50
Vogelsang LB, Marissen R and Schijve J (1981) A new fatigue resistant material: aramid reinforced aluminum laminate (ARALL), Report no. LR-322, Delft University of Technology, Department of Aerospace Engineering
Volesky B, Naja G (2005) Biosorption: application strategies. In: Harrison STL, Rawlings DE, Petersen J (eds) Proceedings of the 16th international biotechnolsymp. cape Town, South Africa, pp 531–542
Wani PA, Zaidi A, Khan AA, Khan MS (2005) Effect of phorate on phosphate solubilization and indole acetic acid (IAA) releasing potentials of rhizospheric microorganisms. Annals Plant Prot Sci 13:139–144
Winge DR, Mehra RK (1990) Host defenses against copper toxicity. Int Rev Exp Pathol 31:47/83
Workman ML, La Charity LA, Kruchko SL (2013) Understanding pharmacology: essentials for medication safety. Elsevier Health Sciences
Wu D, Fang Y (2003) The adsorption behavior of p-hydroxybenzoic acid on a silver-coated filter paper by surface enhanced Raman scattering. J Colloid Interface Sci 265(2):234–238
Yam KL (2009) Encyclopedia of packaging technology. Wiley, New York. ISBN:978-0-470-08704-6
Yokel RA (2000) The toxicology of aluminium in brain: review. Neurotoxicology 21(5):813–828
Zweben C (1992) Metal-matrix composites for electronic packaging. J Min Met Mat S 44(7):15–23

Environmental Implications of Reuse and Recycling of Packaging

Shanthi Radhakrishnan

Abstract A major issue confronting the community and government is waste disposal. A considerable amount of waste is generated when consumers procure and use products. The disposal of waste has become a serious problem due to the increase in the number of consumers and high waste generation. According to the Manual on Municipal Solid Waste Management, Ministry of Urban Development, Government of India (2000), approximately 0.1 million tons of municipal solid waste was generated in India every day leading to almost 36.5 million tons/y. Many manufacturers and retailers are earnest in promoting their brands by the use of excessive packaging without regard to real human needs or environmental concerns. This has resulted in an enormous amount of waste, which calls for numerous waste-management policies. Reduce, reuse, and recycle are tools to minimize the negative implications of manufacturing and retailing on the environment. Design experts today create packaging solutions that consider the optimum use of raw materials and techniques as well as the use of recycled materials in inventory, which can in turn be reused, recycled, composted, or become a source of energy recovery. This chapter presents an overview of the impact of the industrial packaging supply chain on environment, the challenges facing environmentally conscious manufacturing, concepts underlying reuse and recycling of packaging, and the trends in green packaging. The environmental implications of the reuse and recycling of packaging were shown to be beneficial in case studies in terms of raw material, resources, cost, and reduction in landfills. Challenges regarding waste may be different in different countries, but the path to addressing the problems of the waste sector may be common. Evidence from case studies proves that

S. Radhakrishnan (✉)
Department of Fashion Technology, Kumaraguru College of Technology, Coimbatore, Tamil Nadu, India
e-mail: shanradkri@gmail.com

S.S. Muthu (ed.), *Environmental Footprints of Packaging*,
Environmental Footprints and Eco-design of Products and Processes,
DOI 10.1007/978-981-287-913-4_7

ecologically sustainable concepts lead to savings in resources and have a positive effect on the environment giving a competitive advantage to firms. Prevention and reduction of waste at the source and circular global economy will be of primary importance for all countries to bring about radical changes toward greening the waste sector.

Keywords Environment conscious manufacturing · Reuse and Recycling · Green packaging

1 Impact of Industrial Packaging Supply Chain on Environment

1.1 Introduction

Packaging of products has become an area of concern because it occupies an important position in the supply chain with the growing necessity of cost minimization, reduction of environmental impact, and increase in electronic operations. The main purpose of a supply chain is the production, transportation, and distribution of products to consumers. The different levels in the supply chain from supplier to consumer are linked with the movement of packed products, which has a bearing on the design-and-manufacturing process, improved layout, and increased efficiency. The fundamental role of packaging along the supply chain requires connectivity with logistics, marketing, production, and environmental aspects. Logistics calls for easy handling of the packages until it reaches the consumer, whereas marketing demands good appearance and right size of the product to inspire the customer to use it. The production department calculates the minimization of time and cost for the packaging of products, and looks for recyclable packaging and least use of raw materials.

While analyzing environmental issues regarding packaging, sustainable packaging is a main frame that should be considered. To enhance the sustainability factor in packaging, manufacturers should look into the primary details such as whether the product actually needs to be packed and the minimum amount of packaging needed to retain its appearance and quality. The package design process evaluates the mass and volume of packages and determines the optimal use of raw materials. Returnable packages encourage closed-loop logistics, and recycling lays the emphasis on the recovery of primary parts of packages. Packaging waste can be used as refuse-derived fuel and allows for a waste-to-energy principle. In some cases, materials must be disposed of and may end up being incinerated or dumped in a landfill. Many manufacturers have started using recyclable materials, such as plastics, cardboard, and paper, and reusing secondary packaging materials for future shipments.

1.2 Impact of Packaging Raw Materials on the Environment

A wide range of raw materials are used for packaging to serve functions such as protection, sales appeal, product identity and information, consumer convenience, and safety. They are usually applied in three broad categories, namely, (1) primary packaging, which forms part of the basic or first packaging of goods; (2) secondary packaging, which covers larger packs and (3) tertiary packaging, which is the outermost wrappings thus facilitating easy transport. Secondary and tertiary packaging materials are most often similar and can be reused or recyclable. The primary packaging materials usually vary and are damaged thereby making it difficult to reuse or recycle.

The American Society of Testing Materials, ASTM D 996-95 (E01) (Terminology of Packaging and Distribution), defines industrial packaging (tertiary packaging) as a package used for the transportation or storage of commodities, the contents of which are not meant for retail sale without being repackaged. The society also defines transport packaging (secondary packaging) as "packaging intended to contain one or more articles or packages or bulk materials for the purposes of handling and or distribution" (Reusable Industrial Packaging Association (RIPA) 2011). Packaging waste statistics show that on average, every citizen in the 27 European Union member states generated 159 kg of packaging waste in 2011, and this quantity varied between 43 and 216 kg per capita across European countries (Eurostat 2013).

Packaging materials commonly used are a wide range of oil-based polymers, which are nonbiodegradable and difficult to recycle or reuse. The most common types of packaging materials used for primary packaging are paper- and pulp-based materials, glass, and metals such as aluminum or steel and plastics. Approximately 36 % of the total Australian packaging market constitutes paper and cardboard packaging, 30 % plastics, 20 % metals, 10 % glass, with the remaining being other types of packaging. In 2002 to 2003, Australia used 4 million tons of printing, writing and packaging paper and cardboard, and 83 % of the waste paper recycled was used for making packaging and industrial paper. In addition, more than 1 billion milk and juice cartons are used each year in Australia, and >2300 tons of liquid paperboard is recovered through recycling for the use of office paper (Tuckerman 2005).

Paper and cardboard are widely used consumer materials derived from wood, a valuable natural resource. Both chemical and mechanical processes are involved in the conversion of raw materials to paper consuming large amounts of energy and water. It has been estimated that 1 ton of paper consumes 20 fully grown trees, and in 2003 to 2004, Australia used 3,863,000 tons of wood for paper manufacturing, which accounts for an approximate 6.9 % increase from 2002 to 2003. Moreover, approximately 1.9 million tons of paper is sent to landfills each year, which could be recycled instead (Clean up 2009). Approximately 50 % of waste material reaching landfills could be recycled or reused, thus saving natural resources and reducing greenhouse gas emissions. Glass is made from natural materials, namely, sand, soda ash, and limestone with dolomite and feldspar and heated in blast furnaces at temperatures >1500–2730 °F before the glass is cooled. It has been estimated that 2 tons of CO_2 is emitted for manufacturing 1 ton of glass (Hugger 2004). Sand mining, heavy consumption of heat energy, and huge volumes of gas emissions mark the production of glass. Glass is recyclable, and scrap glass called "cullet" is added in different percentages (45–100 %) to the raw material as per the needs and requirements of the end product.

The major global environmental issues with relevance to the plastic industry are numerous. Plastic production uses large amounts of fossil fuel energy and raw material resources with the release of greenhouse gases such as CO_2, N_2O, CFCs, and CH_4 into the atmosphere. Depletion of the ozone layer due to the release of CFCs can cause higher levels of UV-B radiation, thus creating problems with human health and changes in agricultural and marine ecosystems. Gases released by the burning of fossil fuels cause acidification of the environment resulting in impairment of fertility of agricultural soil and damage to crops, forests, and inland fisheries. Urban litter is another problem due to the improper disposal of plastic and paper packaging, thus leading to problems for animals, birds, and marine life.

Raw material for the production of primary aluminum is bauxite, and secondary aluminum comes from aluminum scrap. Molten aluminum is converted into different shapes called "ingots." It is then made into sheets or foils during the semimanufacturing stage, after which the final product is fabricated. Aluminum is a strong metal that can be used as a future resource. The environmental impact of aluminum production is red-mud generation and land use in bauxite mining. Aluminum-associated toxicity through air, water, and soil emissions and wastes from the plant are some of the environmental impacts that cause problems to humans and the environment. The greenhouse gas emissions generated during different stages of aluminum production are given in Table 1. Aluminum toxicity affects approximately 40 % of agricultural soils in the world (Flaten et al. 1996) and has led to fish species extinction due to acid rains. Chronic renal failure, anemia, encephalopathy, and osteomalacia are some of the toxicity effects of aluminum in humans.

The production of steel is accompanied by unwanted products—such as scrap, slag, and scale—that are used by cement and recycling industries; air emissions such as CO_2, CO, SO_x, NO_x; and dust and water emissions, e.g., oil, grease, chemicals, and suspended solids. The impact of steel production is assessed in terms of global warming potential (467 $kgCO_2e$), ecotoxicity (57.3 $PAFm^2y$), fossil fuel (478.4 Mj), carcinogen (DALY 1.2×10^{-5}), and respiratory inorganic (DALY 8.0×10^{-4}) (Tongpool et al. 2010). After aluminum and glass cans, steel cans have a moderate impact on the environment. They are not compostable, but they are recyclable, and are designed for disassembly because they can be taken apart after their end of life for reuse or recycling. Since steel cans are recyclable, they can contribute to reduced resource use.

Table 1 Range of GHG emissions during the different stages of aluminum production (Liu and Muller 2012)

Sl. No.	Production stages	$kg\ CO_2$-eq/kg output				Typical range
		Place/year	Minimum	Place/year	Maximum	
1.	Primary aluminum	Greenfield smelter, Iceland	5.92	Middle East, 2000	41.10	9.7–18.3
2.	Secondary aluminum	Scrap remelting, Europe, 2005	0.32	China	0.74	0.3–0.6
3.	Rolling	82 % yield ratio, US, 1995	0.20	Foil production, Europe, 2005	1.35	0.6–0.9
4.	Extrusion	69 % yield ratio, US, 1995	0.28	International Aluminum Institute, London, 1998	0.74	0.3–0.7
5.	Shape casting	45 % yield ratio, 1995	0.48	International Aluminum Institute, London 1998	0.62	0.5–0.6

1.3 Case Study of the Impact of Packaging Materials on the Environment

A multidimensional environmental evaluation of packaging materials was performed using three methods, namely life-cycle assessment (LCA), analytic hierarchy process (AHP), and cluster analysis (CA) (Huang and Ma 2004). All material requirements, e.g., energy consumed, emissions, waste, and environmental impacts associated with product, process, and service, were quantified. Functional units for the packaging materials and system boundaries were identified. Four stages of the life cycle of the packaging materials were considered: resource extraction, manufacture, use, and waste disposal. SimaPro 4.0 was the software used for computing the LCA. The nine environmental issues considered were ozone-layer depletion, heavy metals, carcinogenic substances, summer smog, winter smog, pesticides, greenhouse effect, acidification, and eutrophication. Under the analytic hierarchy process, seven evaluation factors were considered from ISO 14021 to assess environmental friendliness, and objectives were set. Every packaging material is weighted according to each evaluation factor, and the material is scored using associated evaluation factors; a total score is obtained by summing scores. The cluster-analysis technique used the LCA points and the AHP scores to describe the character of the material and also showed the homogeneity or heterogeneity of the materials.

Table 2 Results of the AHP scores, LCA points, and cluster-analysis grouping of packaging materials (Huang and Ma 2004)

Sl. No.	Packaging materials	AHP scores[a]	Assigned order[b]	LCA points[c]	Assigned order[b]	Cluster-analysis grouping[a]
1.	PET containers	1.32	1	7.75E-04	4	1
2.	PP containers	1.32	1	3.50E-04	3	1
3.	HDPE containers	1.27	3	2.66E-04	1	1
4.	PS containers	1.02	4	8.95E-04	6	1
5.	Steel cans	0.73	5	9.53E-04	7	2
6.	Glass containers	0.63	6	2.69E-04	8	3
7.	Cardboard boxes	0.61	7	5.13E-04	4	2
8.	Liquid paperboards	0.61	7	9.53E-04	2	2
9.	Aluminum cans	0.49	9	2.83E-04	9	3

[a]Higher scores have less environmental impact
[b]In environmentally friendly order
[c]Higher points have more environmental impact

The results of LCA indicated that aluminum cans and glass containers had the greatest impact on the environment when the whole life cycle was taken into account because aluminum cans emit more carcinogens and glass containers emit more heavy metals during the manufacturing stage. LCA also showed that the worst environmental impact (>90 %) was incurred during the manufacturing phase. The LCA scores, which showed higher points, had greater environmental impact. The data for evaluation is given in Table 2. The results of AHP showed that PET, HDPE, and PP containers are better in terms of environmental impact compared with cardboard boxes, liquid paperboards, and aluminum cans. This may be due to the fact that the factor weight assigned for each parameter yielded higher scores e.g., "reduce resource use" had a factor weight of 0.20, and "recovered energy" had a factor weight of 0.13. The higher the scores, the lesser was the impact in the AHP analysis. The cluster analysis formed three groups. Group 1 included all types of plastics because they have low environmental impact and have similar characteristics. Group 2 are cardboard boxes, liquid paperboards, and steel cans because they are "moderate" in terms of environmental damage. Group 3 included aluminum cans and glass containers because they have the worst impact on environment. In the three groups, AHP and LCA scores yielded the same directions in terms of results.

This study provided a holistic approach to analyzing the environmental impact of packaging materials on the environment taking the quantitative (LCA) and qualitative (AHP) methods, as well as the impact on all phases of the life cycle, into account. Materials commonly used in the market were considered, but this study can be extended to other materials.

1.4 Effect of Packaging Supply Chain on the Environment

The packaging supply chain is increasingly complex and involves the active participation of producers of the raw materials using in packaging, the retailers who sell the packed goods, the customers, and the companies who manage end-of-life of packaging. Each member of the supply chain must collaborate and support each other to achieve the best environmental sustainability. The quantity of packaging materials that fill the dustbins show that optimum packaging has not yet been achieved and that redesign of packaging to save raw material and energy is essential.

Optimal packaging design reduces the excessive use of raw material and reduces environmental impacts right from source, manufacture, distribution, and delivery. Communication among all of the partners of the supply chain, as well as the adoption of new innovative technologies, can reduce costs for packaging and distribution, thereby reducing environmental impacts. The packaging supply chain can lower costs by reducing distance and transportation costs, production waste and costs, and unplanned activities to shrink the environmental footprint. Other issues, such as social elements, are also taken into account. The most fitting example would be the interaction of economic consideration with social and environmental issues such as noise pollution, congestion, and CO_2 emissions in packaging and logistics, and their role in retailers' sensitivity to sustainable issues in the supply chain.

The integration of sustainability concepts into legislation will change the environment in which firms work and the nature of competition. This calls for supply-chain managers to address new issues such as reverse supply chain, responsibility for pollution, extent of recycling and reuse, and end-of-life product management. This line of thought will produce changes in existing practices to create new production and management systems.

Our imagined future will call for new renewable resources for packaging and distribution, level of negative impact on the environment, market force and consumerism, attitudinal and lifestyle changes, and policies necessary to achieve sustainability along the entire supply chain. Research focus should be directed toward critical operational and sustainability issues such as decentralization in collection and processing of end-of-life products, better use of used products, and life-cycle analysis interlinked with statistical packages. The closed-loop nature of sustainable supply-chain management will alter the policies and strategies of firms and the competitive environment. Economics will look at total environmental cost, which includes effects on resource depletion and the generation of byproducts such as pollution and waste. Strategy and planning must include sustainability issues, and organizations must go beyond the normal limits to work for a sustainable supply chain, of which packaging and logistics form an important part. Case-study analysis, hypothesis testing, and multiple case-model development are some of the tools that could be used for bringing about awareness in all areas of supply-chain management.

2 Issues in Environmentally Conscious Manufacturing and Product Recovery

A new era of human experience began during the industrial revolution when a shift from hand-made to machine-made products resulted in increased productivity and flow of income; during this period, a higher standard of living transformed the life of the individual and society as a whole. However, industrial production was unplanned and unaccountable, and there were many flaws in the production-and-consumption process leading to indiscriminate use of resources, energy, and materials. Intentional careless approaches in production and consumption have lead to pollution of air, water, and soil and an increasing amount of garbage. During the last decade, environmentally conscious manufacturing and product recovery (ECMPRO) has become the most popular move to safeguard the environment for future generations. Environmentally conscious manufacturing is concerned with manufacturing new products with the utmost care in terms of conceptual design to final delivery and end-of-life disposal of the product such that it matches the specifications prescribed to satisfy environmental standards and requirements. Product recovery, in contrast, is dedicated to minimizing the amount of waste going to landfills through disassembly, reuse, and recycling.

The life-cycle analysis of a product should be well understood to design environmentally friendly products. Life-cycle analysis covers the design development, manufacturing, use, and disposal phases of a product, and ecofriendly decisions are essential at all stages of the product life cycle. "Design for recycling," "design for environment," and "design for disassembly" are some important terms in design development. The end-of-life management of products is crucial and can be achieved by disassembly and recycling. Some meaningful terms regarding environmentally conscious manufacturing are given here (Olson and Sutherland 1994).

Demanufacturing is a process of reducing and retrieving usable components from a product successively through assemblies for major parts; subassemblies for minor parts; and materials for use.

Disassembly is the reduction of a product to its assemblies and subordinate parts. Disassembly is seen as a cost-adding step to demanufacturing.

Rebuilding returns a product to an as-was condition. It implies that the product is minimally refurbished where essential worn parts are replaced.

Recycling is a process of extracting useful materials from waste by sorting, mechanical, or chemical operations.

Many environmental regulations have been passed by countries around the globe for extending the responsibility of product end-of-life management to manufacturers. This requires considerable focus in understanding how the energy and material flow is affected by product changes and how to plan desirable changes in the industrial ecosystems. These changes will have an impact on material choices, energy use, market and consumer response, waste management, and policy changes. Current environmental regulation is centered on environmental health and safety, but it can extend to other criteria such as social and humanitarian grounds. This creates a great number of challenges to manufacturers and retailers who strive hard to sustain their position in the competitive global market.

2.1 Challenges Facing Environmentally Conscious Manufacturing

2.1.1 Manufacturing Challenge

New laws have been created to give a basis to define labels that state "clean," "green," "ecofriendly," etc. Manufacturers must meet these changing definitions and choose materials that have less impact on environment, minimize materials for packaging and other uses, and develop efficient recycling schemes. Environmental regulatory compliance is seen as the minimum standard of performance, and consumers today are asking for more far-reaching benefits. It is best to integrate environmental considerations into corporate culture and business planning. The main goal of the manufacturer should be reduction of negative environmental impact in all subsystems to achieve an overall rating of the system as "ecofriendly." Many organizations have reported savings by minimizing packaging and type of raw material used. Firms have switched from white to brown boxes saving material cost as well decreasing the bleaching of paper in the process sequence. Similarly, environmental impacts must be identified in all areas of manufacture and distribution to improve their environmental performance.

2.1.2 Role of the Consumer

The consumer is very important and meeting his or her needs is the primary aim of the manufacturer. The performance, quality, and cost of the product must suit the customer's needs as well as satisfy environmental compliance guidelines. The attitudes and values of consumers has changed over the years, and they have become aware of their role in reducing the environmental impact. A study reports that in 1989, 67 % Americans were willing to pay 5–10 % more for green products; by 1991, environmentally conscious people were willing to pay 15–20 % more for ecofriendly products. A survey in 1994 showed that 79 % of female consumers surveyed in the United Kingdom were willing to pay 40 % more for a product that has been proven to be green (Laroche et al. 2001). The attitudes of the consumer is changing: They want authentic certifications for products that have been developed from recycled materials or for the use of recycled materials in their packaging. Consumer forums have publicly criticized Walmart and Procter & Gamble for putting a green label on their brand of paper towels that were chlorine bleached and made of virgin material and packaged in plastic; the claim made by the manufacturers were that the inner tube for the towels was made from recycled paper. An effort made by McDonalds to eliminate polystyrene clamshell packaging was commended as progressive and exemplary in corporate practices regarding environment. This clearly shows that consumers refuse to be influenced by false environmental claims.

2.1.3 Design Consideration

The potential impact of the life cycle of a product should be minimized. Green quality function deployment (QFD) considers product quality requirements, environmental impact, and production costs. Further developments in green designing also include life-cycle analysis, life-cycle costing, and analytic hierarchy process. Life-cycle engineering is an LCA-assisted method where the product life cycle is designed by making choices on product concept, structure, materials, and processes (Ilgin and Gupta 2010). Material selection covers factors such as weight, processability, and cost; material selection charts integrate environmental concerns, green material cost analysis recommends materials that cause less pollution. The recyclability of selected materials is an important factor, and the recyclability index is a tool for evaluating the material recovery. Another important concern is reverse and closed-loop supply chains, which are involved in the collection and recovery or disposal of used products. While designing products, uncertain characteristics should be included to prevent design problems and to allocate alternatives for the foreseen problems in the design. Design evaluation by simulation techniques serves to examine the impact of the design on long-term basis of a closed-loop supply chain with recycling activities.

The end-of-life option for a product must be determined in the design stage. Recovery and disposal are the options, but the basis of selection is based on environmental impact, legislation, quality, and cost. The next problem is marketing of developed products, and the issues include pricing manufactured and remanufactured products, competition involved in remanufacturing, and determining the return policy. Product design plays a crucial role in terms of environmental impact, and care must be taken to spell out all consequences at the designing stage. Henry Ford highlights the importance of optimal design by the following words.

> Waste is not something which comes after the fact… Picking up and reclaiming the scrap left over after production is a public service, but planning so that there will be no scrap is a higher public service.

2.1.4 Evaluation of Environmental Impact

Measurement of environmental impact is highly associated with life-cycle analysis. In earlier days product designers and chemical-process designers were primarily concerned with the life cycle from raw material selection to the manufacturing and product-completion stage, but currently there is environmental concern at all stages of the product life cycle. Process engineers must clearly understand their product process sequence, as well as the byproducts that may be formed, and find solutions for the use or disposal of the same, i.e., the manufacture of vinyl chloride is associated with the byproduct generation of hydrochloric acid, which can be used for steel or semiconductor manufacturing (Allen and Shonnard 2001). The environmental impact of different processes is taken into consideration, and the best route is usually selected as shown in Table 3.

Table 3 shows two alternative synthesis routes for the production of methyl methacrylate and their environmental implications. In this case, the health and safety issue associated with sulphuric acid is the major concern and hence the isobutylene route is preferred. Although more data are available for the two processes, the required information is taken according to the needs of the industry.

Table 3 Stoichiometric, persistence, toxicity and bioaccumulation data for two synthesis routes for methyl methacrylate (Allen and Shonnard 2001)

Compound	Lb (kg) produced or required per lb of methyl methacrylate[a]	Atmospheric half-life/aquatic half-life[b]	l/TLV[d] $(ppm)^{-1}$	Bioconcentration factor[e] (concentration in lipids/water)
Acetone-cyanohydrin route				
Acetone	−0.68 (−0.31)	52 days/week	1/750	3.2
Hydrogen cyanide	−0.32 (−0.15)	1 year/week	1/10	3.2
Methanol	−0.37 (−0.168)	17 days/days	1/200	3.2
Sulphuric acid[c]	−1.63 (−0.74)	–	1/2 (est.)	
Methyl methacrylate	1.00 (0.45)	7 h/week	1/100	2.3
Isobutylene route				
Isobutylene	−1.12 (−0.51)	2.5 h/week	1/200	12.6
Methanol	−0.38 (−0.172)	17 days/days	1/200	3.2
Pentane	−0.03 (−0.014)	2.6 days/days	1/600	81
Sulphuric acid[c]	−0.01 (−0.005)		1/2 (est.)	
Methyl methacrylate	1.00 (0.45)	7 h/week	1/100	2.3

[a]A (−) stoichoimetric index indicates material is consumed, whereas a (+) index indicates a product of reaction
[b]Atmospheric half-life based on hydroxyl radical reaction; aquatic half-life calculated by way of biodegradation based on expert estimates
[c]Lifetime of H_2SO_4 in atmosphere is short due to reactions with ammonia
[d]TLV is the threshold limit value, and the inverse is a measure of inhalation toxicity potential for a chemical
[e]Bioconcentration factor is the chemical's potential to accumulate through the food chain

LCA analysis has several shortcomings. A LCA is data- and resource-intensive and tends to include everything, which may result in false impressions: (1) It is complex in nature in both methodological and analytical terms; (2) it takes a limited input/output approach and neglects qualitative or nonquantifiable variables and uses inadequate substitutes for environmental impacts; (3) it is directive in nature and cannot provide concrete measure for the greenness of a product; and (4) it is sometimes very confusing with the methods and data available; experienced personnel and collective analysis are necessary to avoid any improper decisions. Studies on the need of reusable versus disposable nappies are still under analysis due to different conclusions. Designers, product engineers, and process engineers are being asked to develop environmentally friendly products without guidance on what is "environmentally preferable" in practical terms, how they can be identified, what are the upper and lower limits, impact on ecofriendly choices on the other parts of the industrial systems, and the entire supply chain.

Further several LCA methods are available, and the choice of the right life-cycle impact analysis (LCIA) can lead to differences in final conclusions. The impacts from emission depends on the quantity of the substance emitted, the properties of the substance, and the characteristics of the emitting source and the receiving environment (Finnveden et al. 2009). A global default procedure is followed in LCIA and will cater to the first two aspects of the impacts per the emissions assessment. However, the situations can be different in terms of locality or region, and hence the same emission quantity may lead to different levels of impact. Therefore, site-dependant characterization is essential. Resources may be of two types—abiotic and biotic—and most of the environmental impacts have been devised to measure biotic resource depletion; similarly land use, water use, and toxicity in indoor and outdoor air require a great degree of differentiation and technical limitations such as mid-value, end-value, etc., to compute the impact of the product, process, or service. The uncertainties of LCIA can be in data (e.g., variability, specification mistakes, incomplete and irrelevant information), in choices (may be inconsistent with the goal and scope of analysis or across alternatives), or in relationships (directions of relations may be wrong, incomplete, hasty, and implemented inaccurately).

To study the environmental impact of products and services, organizations use various tools for evaluation to prevent any flaws in the approach. A research report highlights the use of environmentally conscious business practices such as Design for the Environment (DFE), life-cycle analysis (LCA), total quality environmental management (TQEM), green supply-chain management (GSC), and ISO 14000 EMS requirements. The subcomponents of each of the above are specified in Table 4. Analytical network process (ANP) or analytic hierarchy process (AHP) are models used for the decision-making framework.

Another study describes three tools, namely, life-cycle assessment (LCA), quality function deployment for environment (QFDE), and theory of inventive problem solving (TRIZ) (Sakao 2007). This combined strategy helps product designers in

Table 4 Components and subcomponents of major environmentally conscious business practices (Sarkis 1998)

Sl. No.	Environment conscious business practices	
	Components	Subcomponents
1.	Design for Environment (DFE)	Design for recyclability (RECY) Design for reuse (REUSE) Design for remanufacturability (REMAN) Design for disassembly (DISASS) Design for disposal (DISP)
2.	Life-cycle analysis (LCA)	Inventory analysis (INVAN) Life-cycle costing (LCC) Impact analysis (IMPAN) Improvement analysis (IMPVAN)
3.	Total quality environmental management (TQEM)	Leadership (LEADER) Strategic environmental quality planning (SEQP) Environmental quality–management systems (EQMS) Human resource development (HRD) Stakeholder emphasis (STAKE) Environmental measurements (EMEAS) Environmental quality assurance (EQA)
4.	Green supply-chain management (GSC)	Inbound logistics/procurement (INBD) Materials managaement (MTMAN) Outbound logistics/transportation (OUTBD) Packaging (PACK) Reverse logistics (REVLOG)
5.	ISO 14000 EMS requirements	Environmental policy (EP) Planning (PLAN) Implementation and operation (IO) Checking and corrective action (CCA) Management review (MREV)

a multifold manner, e.g., the designers could use the LCA results to identify that the product had high impact on global warming through energy consumption; this could be remedied by defining a requirement in QDFE to reduce energy consumption; TRIZ allows designers to generate four solutions for improvement of which one is used based on the requirement. This methodology has more benefits than the independent use of the three tools.

2.1.5 Product Recovery

Product recovery is an integral part of converting waste into resource. The cooperative effort of consumers, retailers, and manufacturers is essential to manage the end-of-life essentials of a product. Many people consider waste as inferior, but waste could also serve as a resource if handled efficiently, e.g., (Richards 1994) a waste stream of acetonitrile was cleaned by British Petroleum and serves as a feedstock for the production of insulin. Instead of disposal of waste, minimizing waste and the reuse of waste could yield efficient results. Similarly Dow Chemicals formed a new business group, Advanced Cleaning Systems, in 1990. The organization was threatened by regulations on ozone-depleting chemicals and the control of toxic air emissions. Apart from developing alternative chemicals and processes, they offer new packs to their customers with a take-back policy of the used chemicals in reusable containers.

Product disposal considerations require efficient end-of-life management. Omission of toxic materials would be ideal for a product that is to be discarded; material diversity should be avoided if the product is to be recycled. Identification marks in different components to show that they are made of the same material would help in the recycling process. Modular compartments can be planned in products that can be easily changed to make them as good as new. System-performance improvements would yield better economics than dealing only with product performance. The challenge for the managerial side would be to ensure that all of the players of the supply chain are environmentally conscious and require managing risk and customer satisfaction at all levels.

3 Reuse and Recycling of Packaging

Delivering products to consumers by preserving their integrity and usefulness is the primary role of packaging. The environmental impacts of packaging have increased considerably and litter ends up in landfills through the municipal waste system. However, the regulating legislations issued by different governments of developed and developing countries have created awareness among manufacturers and consumers to take into account the environmental footprint of products. Raw materials cost has increased, and the impact of packaging on the total cost is very high; this calls for sustainable packaging management. Retailers are imposing certain requirements on suppliers to manage environmental impacts. Walmart has established a comprehensive packaging score card on which suppliers are evaluated. Preconsumer and postconsumer packaging chains are analyzed and designed in such a way that recycling or reuse has the lowest environmental impact or the waste is brought back and recycled to prevent waste from being sent to landfills.

Many life-cycle assessments are being performed to assess the environmental impacts of packaging; the effect of reuse and recycling are also assessed to find the best solution for decision making in terms of design and product development. LCA can be performed as cradle-to-grave, cradle-to-gate, or gate-to gate, but environmental legislations calls for assessments along the entire life cycle of the product until it is reused or recycled with minimal environmental effects. The circular economy has been recommended for good results. It has two characteristics, namely, the biological flow designed to safely re-enter the biosphere and the technical flow, that are designed to circulate at high quality without entering the biosphere. This is not the same as with the current linear economy, which uses natural resources and materials without any concern for the ensuing environmental impacts. In the circular economy, there is no waste because all of the biological and technical components are designed to fit back into the natural cycle. Long-lasting products, diverse and versatile components, and products that can be upgraded or repaired should be the primary focus in product manufacturing. Obtaining energy from renewable sources and thinking in terms of systems are important for reducing the environmental impact.

3.1 Concepts Underlying Environmental Impact Reduction

The reduction of environmental effects can be addressed by using the principles of reduce, renew, reuse, and recycle. Reduction of the use of materials can be achieved in containers and packaging by reducing product weight, making the walls of the container thinner, achieving compactness in design of the product pack or container, and downsizing products. Attempts in introducing renewable packaging, such as polyactide or biopolyethylene, instead of petroleum-based products are important. Reuse can call for the development of refills and replacement products so that the original packs or containers can be reused. Recycling initiatives involve the use of recycled materials such as recycled paper or recycled resin. It also recommends the after-the end-of-life product should be recyclable or biodegradable with minimal impact on environment.

3.2 Implications of Reuse

Reuse and recycling are the focus, and this has brought a new insight and awareness of the amount of waste that is generated during product production and consumption. During the European week for waste reduction, 79 reuse centers in 5 EU countries collected 709 tons of goods, of which 129 tons have been reused (Reuse 2014); otherwise it would have gone into the waste cycle. Reuse targets are to become an important part of the waste legislation in the EU. The KAO group in Japan (KAO 2014) undertook a design transformation in their laundry detergent bottle by reducing the amount of resin per bottle, making the walls thinner by 29 % leading to reductions in environmental impact by approximately 2800 intones of CO_2/y and a reduction in costs by approximately 350 million yen. The company also used 10 % biopolyethylene for shampoo refill packs and reduced CO_2 emissions by 12 %; a shrink film containing 50 % polylactide was used for their green tea packs resulting in a 38 % reduction, and they are also aiming at changing the packaging for other products as well.

In earlier days, the reuse of postconsumer packaging, such as glass bottles and jars, was common. In cases where production was in a central facility and distribution and collection points were far, leading to transport expenditure, reuse was the principle used. Refillable bottles are stronger and can be reused, thereby generating less packaging waste than single-use containers. It has been estimated that 46 kg of aluminum is necessary to fill 1000 l of beer, whereas the same amount of beer needs 26 kg of glass, thereby reducing the use of resources by 43 % (Mehr 2015). Furthermore, glass bottles can be reused. Curbside-collection schemes under the Green Dot systems show lower collection rates and recycling rates and percentage compared with the deposit schemes. In deposit schemes, as in reverse vending machines or incentives for waste returns, 99 % of materials are recycled, and here bottle-to-bottle recycling can be performed (PricewaterhouseCoopers 2011). Refilling and take-back schemes are available but only as activities performed by local businesses rather than as large-scale activities. The trend is moving back to reuse for environmental concerns where reusable bags are slowly replacing single-use carrier bags, and levies and bans on lightweight carrier bags seem to turn the attention of the consumer toward reuse.

3.3 Issues Related to Recycling

Recycling is always associated with recovery, and waste management is very important for recycling to achieve viable solutions. PET is the most recycled plastic material in Europe, and it was estimated that approximately 65 billion bottles were recycled in 2013 (PETCORE 2014). Another report states that huge investments are being directed toward the development of PET-reclamation plants to increase recycled content in new bottles. PET-reclamation capacity is expected to increase by 50.3 % in the next 3 years (Powell 2011), and 12 of the current 20 PET-reclamation plants in the United States and Canada produce RPET for the manufacture of new containers. The ability of PET to be recycled and reused for different end uses is helping in achieving a circular economy. Any waste can pass through multiple stages, such as manufacture into a reusable container, enter into the waste stream, recover for energy, and recycle into a durable application.

Landfill: Space for landfills is becoming less, and it is advisable to use this option as the last stage in waste management. Well-managed landfill sites can result in limited environmental harm apart from collection and transport; the long-term risks of waste are the contamination of soil and groundwater due to the breakdown of substances in the waste to become pollutants. Once a waste material reaches the landfill, it has passed all stages of reuse and recycling and cannot be recovered, thus showing that the material flow in landfills is linear. Diversion of waste from landfills are the primary concern, and many governments have passed tough legislation to prevent waste from reaching landfills.

Incineration and energy recovery: Energy recovery is the main concern in many countries that incinerate plastic waste, which can lead to the release of hazardous substances into the atmosphere. In countries such as Denmark and Sweden, the infrastructure for incineration is extensive, and this technique is used to deal with municipal solid waste. The energy recovery may be varied because it can be used for electricity generation, combined with heat and power, or used as solid refuse fuel used in blast furnaces and cement kilns.

Downguaging: Many manufacturers use the required material for a given application. However, for the sake of aesthetics, convenience, and marketing benefits, an overuse of packaging results. Existing investment in tooling and production processes can also result in the excessive packaging of some products. The principle of reduction in the amount of packaging per unit will help in reduce waste volumes.

Levels of Recycling: Recycling may be performed as four levels: primary, secondary, tertiary, and quaternary. Mechanical reprocessing of a product into a one with similar properties is primary recycling, also known as closed-loop recycling. In secondary recycling, there is mechanical reprocessing of a product into a product with lower-grade properties. Recovery of chemical constituents is tertiary recycling, also described as chemical or feed stock recycling, e.g., depolymerization of a polymer into its chemical constituents. Quaternary recycling is energy recovery, energy from waste, or production of energy from the decomposition of waste as biogas or by biological treatment with anaerobic digestion. A report (EGF 2015) states that 2000 tons of waste produce energy to supply 150 thousand inhabitants. After treatment, the bottom ashes may be used for civil construction, public works, and landscape recovery. Many ferrous and nonferrous metals are recovered from the combustion process and sent to recycling industries.

Reuse can save raw materials and energy, but care must be taken to see that the packaging materials are fit to be reusable. Primary packaging may not be reusable, but all other types can be reused in areas where the multiple layers are wrapped before being put into the outermost packaging. Tertiary packaging can be used for secondary packaging and so on. It is in the mindset of the organization to use packaging materials effectively to bring savings in cost, raw materials, energy, and transport. The trend of reuse is more common in developing countries than in developed nations. The material is reused until it becomes unfit for reuse, at which point it is sent for recycling. When reusing materials, this should be performed so smartly that one must be unable to recognize that the packaging has been reused. The cost-effectiveness of refills also encourage the reuse of packaging. Recycling is the last stage after reuse when the packaging is considered as waste to be developed into a raw material for a new product. Recycling attempts are developing into large industries with huge investments and will serve to encourage manufacturers to use recycled materials. The range of materials obtained from recycling activities include energy, ashes for landfill, organic compost, recycled raw materials, and refused derived fuel. Apart from environmental benefits, recycling and waste management can offer employment opportunities for many ranging from grassroots-level personnel to researchers and ecodesigners. Consumers today are ecoconscious and tend to purchase products with recycling labels and recycled content. The environmental safety trend has begun and will tend to protect the environment for the future.

4 Trends in Green Packaging

In this age of INTERNET marketing, sustainable packaging is a yardstick of competitive stature. Ecofriendly packaging puts manufacturers in a better marketing position and showcases their product as a quality and environmentally conscious product. Sustainable packaging has become a part of day-to-day activities and is just another requirement such as product performance, service, and pricing. Governments, producers, retailers, and consumers have started realizing the importance of green packaging and its importance on our environment, yet there is no organizational framework that can be adopted by all manufacturers. The projected market for sustainable packaging will be $244 billion by 2018. Companies who are progressive in their approach will use the concept of sustainability in packaging to surpass their competitors and differentiate themselves from other manufacturers in the marketplace.

Green packaging is of great importance to humans and their environment. Packaging production has always used fossil fuels, which adds millions of metric tons of carbon dioxide and methane into the atmosphere, whereas discarded packaging ends up in landfills and oceans causing soil, water, and plant contamination, which then pollutes the food chain. Sustainable packaging can eliminate these contaminants by lowering packaging content, formulation of recyclable or biodegradable packaging, and use of wind, solar and biofuels for the production and transport of packaging. It can be anticipated that packaging is the main source driving the green market and is one of the most important contributors to the demand of going green. There are various trends surrounding green packaging, and a few have been discussed here.

4.1 Green Consumers

The primary motives that drive a consumer in terms of regarding environmental impact and sustainability are helpful to know when creating new product positions. Consumers have been classified (Iyer and Banerjee 1993) as "planet passionates" who want to preserve their planet by recycling bottles, cans, and newspapers; the second group is "health fanatics" with the motive of preserving personal health through the use of safe food and organic products only; the third group is "animal lovers" with the motive of preserving animal life by becoming members of humane societies, buying cruelty-free cosmetics, and boycotting fur coats and leather goods. The growth of green products is based on such studies, and this motivates green consumers to buy/avoid such products. It has also been found that green advertising is centered on planet preservation and environmental issues and is generally focused on the producer and production toward sustainability.

4.2 Sustainability Measures

The latest trend is the shift from petroleum-derived to plant-based plastics. Leading manufacturers and retailers look out for biobased raw materials, and they tell their consumers about the savings in fossil fuels due to the use of ecofriendly packaging materials. Many global brands have jointly established a Bioplastic Feedstock Alliance (BFA) to increase awareness and set standards for the wider use of plant-based plastics.

Paper and board has been estimated to increase to 6 % by 2017, which will amount to >30 million tons with an approximate value of $70 billion (Lifshitz 2014a, b). Brands will seek out for new barrier technologies that are renewably sourced, recyclable, and biodegradable. Bioplastics and water-based coatings are some of the barrier technologies that will exist in the near future.

4.3 Lightweight Packaging

Lightweight small-sized packaging with little input of raw materials and environmentally friendly rigid packaging is a central theme. The evolution of a light weight packaging standard, known as "packaging-efficiency model," provides a holistic approach covering the packaging life cycle. The prime concerns are resources used, protection of product, efficiency in transport and delivery, and provision of a positive customer experience. Problems associated with opening of packaging will be addressed by providing frustration-free packaging where packaging can be easily opened without tools.

4.4 Consumer Information

In the United Kingdom, it has been estimated that 5.91 million tons of packaging waste (Wadhwani 2014) accumulates every year. Consumers want answers as to the source of the packaging material, the materials used for packaging, and the possibility to recycle the package. Traceability is both an environmental and ethical issue. Today brands are communicating their sourcing details and supply-chain traceability to assure their consumers that their packaging is from legal and sustainably managed production cycle.

A universal on-the-pack label should be made mandatory for the consumer to understand sustainability information without confusion regarding the level of sustainability in the product. Regarding recycling labels, the method of recycling should be clearly given to enable the customer to understand the method adopted for recycling and its impact.

Reusability and secondary use in packaging also attracts the consumer's interest. In 2011, Pizza Hut introduced a special Green Box (Lifshitz 2014a, b) that could be broken down into four plates and a small box for leftovers. These endeavors create a strong impression on the minds of the consumers, and they become willing to pay more for such innovative sustainable ideas.

4.5 Consumer Attitudes

The boom of the health and wellness sector has become a key factor in the packaging trend showing a wide public desire to purchase what is environmentally sound and sustainable. Health credentials are to be displayed on food-related packaging. The special features of the product in terms of natural-ingredient formulations, innovative methods of preservation of fresh food, and communicative labeling will help in both short- and long-term success of the product and its manufacturer.

Another important tool for brand managers is advertisement about the product. Advertising on packaging increases the awareness of credibility and authenticity and reassures the customer about the truthfulness and high quality of the product. The brand owner can further capture the attention of the consumer by communicating the carbon-footprint benefits of using a local brand because transportation and packaging will be reduced.

Although consumers are willing to pay more for sustainable products, it is the responsibility of the manufacturer or retailer to offer the best cost-effective price because cost is the first consideration in buying decisions. Buying in excess and stocking has disappeared, and consumers today are looking for sustainable reusable packages for consumption options.

4.6 Package Design and Brand Imagery

Package design is an important aspect in creating a brand image. Packaging assists consumers to select among many of products offered on a retail shelf. "Badge products"—such as cigarettes—use packaging as an advertising tool. The packaging has a high degree of social visibility and is constantly taken out and left out for public display. These products also have the highest brand loyalty and show the association of the consumer with the brand image. Health warnings are discreetly incorporated into the packaging design to minimize their intrusiveness and preserve the look of the packaged product. Although cigarettes are injurious to health, package design plays an important role in serving as a social status symbol for the user.

4.7 Green Product Innovation

Product innovation in terms of sustainability is moving toward higher goals for recycled content and recyclability in sustainable packaging. Starbucks recently announced a new goal to have 100 % recyclable or reusable cups, and Hewlett-Packard cut packaging on a few products by 97 % (Maria and Pujari 2010). Water is being sold in a 100 % recyclable paper carton instead of the traditional glass bottle and has the label "Boxed Water is Better for Earth" (Nobel et al. 2009). Form, fill, and seal is used for the manufacture of this product given that the main marketability of the product is due to the recyclability of the container.

Earlier manufacturing worked on the principle that solutions to malfunctions or breakdowns were replacement of the entire assembly or subassembly. Manufacturers are developing designs that avoid hazardous components and bring savings by the concept of reuse. Modular designs help in principles of remanufacturing, automated remedies for problems and repair, or replacement by subvendors or manufacturers. Careful material planning may result in losses upstream, which could be compensated by savings downstream and in recycling. In the United States, disposal costs have reached $50 billion (Kleindorfer et al. 2005) with no clear solution of how costs can be covered. Waste can also contain toxic parts; disassembly can help to remove the toxic parts, and the rest could be sent to the landfill. Promoting environmental care, mitigating environmental health, and the safety impacts of a company are the basics of socially responsible good business.

Green-product innovation starts with the integration of environmental concerns and conventional product attributes at an economical price. Usually development and manufacturing costs are high, thus making the price noncompetitive, which slows down the introduction of new products. Another challenge is the lack of awareness about green products among consumers, but this can be rectified by the use of ecolabeling and third-party certification. Organizational and managerial issues are important when dealing with third-party certification and require environmental specialists to work in the product design and developmental process. An understanding of a green-product portfolio will enlighten how companies invest in the green-product technologies platform to bring new green products to markets.

4.8 Design-Based Research

Design experiments combine design focus and assessment of critical design elements. Qualitative measures assess the performance of the design in practice and determine how social and contextual variables affect cognitive variables. Design-based research is essential to produce green designs with a special focus on environment. This calls for creative thinking and problem-solving approaches. Successful implementation of designs call for many experiments in the practical context. The egg-drop experiment (Dede 2005) is part of a research activity where the researchers are given raw eggs and a few basic materials. The researchers are asked to create packaging that will cushion the egg from breakage when dropped from a considerable height. Similarly, research in design is required for environmental solutions, and manufacturers must identify the most optimal green design with end-of-life management that will give the least impact to environment. Thus, researchers will be able to identify the best solutions, and when these strategies are implemented they will not be met with impractical failures.

4.9 Sustainable Product-Design Tools

The design phase of any product is attributed to 5–7 % cost, but the decisions made in terms of product design will cover 70–80 % of the total product cost. In the early design phase, a high level of uncertainty prevails regarding the working ability and sustainability of the given product or process. Novel methods and tools are necessary to help designers ascertain this aspect. DFE enables environmental consideration coupled with business opportunities, and the standard should be set by the organization for the level of DFE to be implemented by the company. Ecodesign tools may be based on checklists, life-cycle assessments (LAC), or quality function deployment (QFD). Check lists are tools with many questions to be answered and are suitable for small- and medium-sized industries. They are subjective in nature and require great knowledge, skill, and experience to interpret the results. The LCA is an objective method of assessment that calculates the environmental impact of the product throughout is life cycle. LCA is not design-oriented, but it is designed to analyze certain structures and components as well as the greenness of concept description. Tools based on QFD convert customer needs into engineering characteristics and conduct a correlation analysis between customer needs and product quality characteristics. The common problem in this method is that correlation analysis is based on the traditional environmental-engineering regulations without considering the whole product life-cycle. The tools are used in combination to enable decision makers to pursue sustainable product-design activities on a holistic basis.

4.10 Sustainable Materials

The use of locally available materials and labor can reduce the environmental costs of shipping, fuel consumption, and CO_2 emissions generated from transportation. Reclaimed and recycled materials can be used for new-product development. Water recycling systems, such as rain-water harvesting and the reuse of "grey water" from households and off-grid homes—which harness energy from active (solar power, biomass, wind turbines, and geothermal energy) or passive (bioclimatic buildings, green roof designs) energy systems—are some of the applications of systems leading to eco materials. Ecodesigners can take an environmentally conscious approach, from the choice of materials to the type of materials being consumed to the disposal of waste, to obtain maximum benefit out of the materials.

The term "green" signifies various meanings, and in the context of sustainability there is no limit to achieving better prospects for the environment. Sustainable packaging must promote the healthy coexistence of humanity and nature, designers accepting responsibility for the consequences of design, the creation of safe products of long-term value, and an understanding of the limitations of design to seek continuous improvement in terms of environmental implications and sustainable concepts. Understanding green-packaging trends, as well as integrating elements such as packaging sustainability with convenience, will increase brand loyalty, enhance product reputation, and bring improvement in business.

5 Roadmap for Green-Packaging Solutions

The roadmap for green packaging solutions and better waste management is given in Fig. 1. It shows the forces that have compelled a change in product and manufacturing systems to work with great concern to reduce the negative impacts on the environment. Waste is being produced at an alarming rate causing massive damage to the ecosystems in the environment. The main statistical findings by Eurostat shows that in 2013, municipal waste generation ranged from 272 kg/capita in Romania to 747 kg/capita in Denmark (Eurostat 2015). Environmental and human health hazards resulted in changes in the global level leading to regulations and laws to reduce waste and to make manufacturing industries and retailers accountable for the waste produced. Many leading brands have used sustainability and ecofriendliness as a tool to gain competitive advantage over other brands, thus driving consumer demand for more green products made from materials that have a low impact on environment. These driving forces have brought a wave of change across the globe to work for green solutions to achieve zero waste throughout the life cycle of the product.

Efforts are to be directed right from the grassroots level to the highest level to achieve maximum benefit. The roadmap highlights the involvement of all stakeholders starting from the consumer, industry, nation, and globe. Attitudinal and behavioral change on the part of the consumer to make environmentally conscious choices and to look for quality and function rather than aesthetics would serve to force manufacturers work toward the goal of sustainability. On one hand, today's consumers look for products with sustainable materials and increased recycled components to help reduce the negative impact on environment. On the other

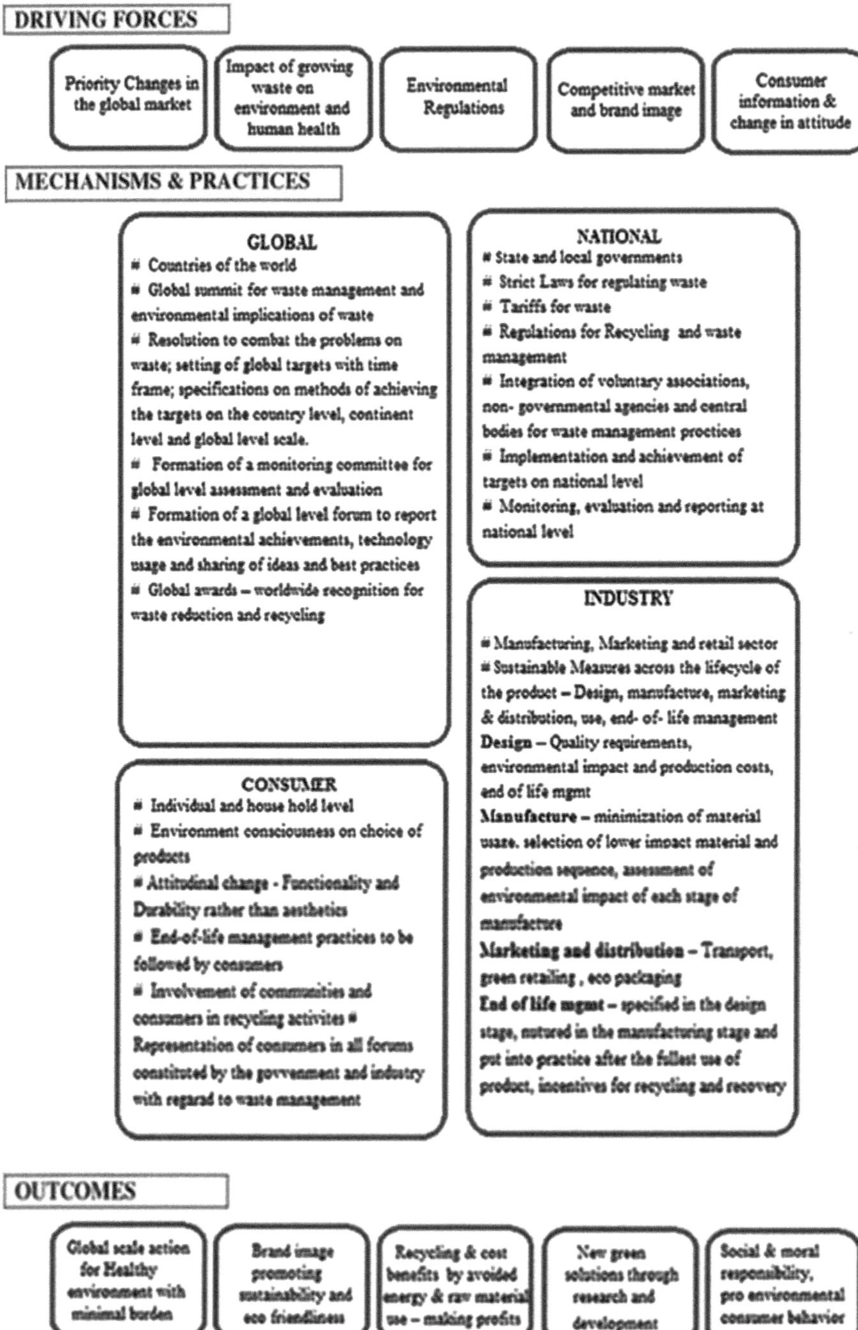

Fig. 1 Roadmap for green-packaging solutions

hand, the packaging industry is working toward ecofriendliness from the design stage to end-of-life management. Reuse and recycling has become very important to showcase the green brand image. To quote a few examples—

1. Leftover sound-absorbing material from the production of cars and sedans has been used to insulate coats that are transformed into sleeping bags for many homeless people. This effort prevented 212,500 pounds of waste going to landfills (General motors 2012).
2. General motors has used oil-soaked booms for under-the-hood car parts and recycled >100,000 pounds of plastic resin along with the use of resources, such as oil and water from the booms, thus saving 29,000 gallons of water and oil and eliminating 212,500 pounds of waste and 149 tons of CO_2 emissions (Price 2010; General Motors 2011).
3. General Motors donated 100 steel crates used for shipping engines and 250 crates from other departments for the conversion of vacant parking lots in Detroit into urban gardens involving the community residents and volunteers to water and maintain the gardens for free vegetables and herbs (General Motors 2013).

These efforts, and many others adopting reuse and recycling, have fetched General Motors the Top Project of the Year Award in the Environmental Leader Product and the Projects Award for driving a global movement for zero waste (Fast Lane 2014).

The movement toward sustainability and green packaging has been further enhanced by the role played by countries and nations around the globe. Several laws and regulations—such as the PlasTax in Ireland, the German Packaging Ordinance, the ban on plastic bags in Africa, and the fee for the use of plastic bags in supermarkets in Hong Kong—have led to a reduction in the use of plastic bags. Targets and time frames for the reduction of waste can be set at national and global levels, and monitoring and evaluation committees could record the progress toward the goal of zero waste. Achievements and savings, best practices, and technologies could be transferred by forums for others to follow the path of sustainability and to show that such measures are feasible. Conscious efforts and policies will slowly lead to a healthy environment where reuse and recycling are performed to create zero waste, thus leading to savings in energy, raw material use, and reduction in emissions resulting in healthy profits. Recycling and reuse could open the doors to innumerous research projects, and new solutions could be developed. Social and moral responsibility on the part of industry and consumers would result in a safe environment where brand image is established based on the extent of product sustainability.

6 Concluding Remarks

Manufacturers and retailers ought to look for functional products with good quality and assign less importance to aesthetics, packaging, and finishing. Intelligent packaging solutions are far more serviceable and sustainable than ones that contribute to pollution of the environment. More involvement is required from all stakeholders, along with active participation of consumers, to take responsibility for waste. The aim of ecodesigners is to focus on the goal of zero waste to achieve the best results; when minor quantities of waste occur, it must be reused or recycled to form the closed-loop system. Efforts to clean up the environment have begun, and awareness for green-product design and process will pave the way to control the generation of waste. Whatever waste is generated will be reused or recycled to produce a circular economy with growth and development. Waste is no longer rubbish but a means for recovery and recycling for new product development. The attitude of the people requires a positive change toward nature and the environment, and the technological march toward sustainability will blossom into a greener environment. To conclude with the words of Evo Morales 'Sooner or later, we will have to recognize that the Earth has rights, too, to live without pollution. What mankind must know is that human beings cannot live without Mother Earth, but the planet can live without humans.'

References

Allen DT, Shonnard DR (2001) Green engineering: environmentally conscious design of chemical processes and products. AIChE J 47:1906–1910. http://library.certh.gr/libfiles/PDF/GEN-PAPYR-4713-GREEN-ENG-by-ALLEN-in-AICHE-J-VOL-47-ISS-9-PP-1906-1910-Y-2001.pdf. Accessed 24 May 2015

Clean up (2009) Paper and Cardboard Fact Sheet. Sep. http://www.cleanup.org.au/PDF/au/cua_paperandcardboard_fact_sheet.pdf. Accessed 21 May 2015

Dede C (2005). Why design-based research is both important and difficult. Educ Technol 45:5–8. http://isites.harvard.edu/fs/docs/icb.topic786630.files/DedeDBR_introEdTech2.pdf. Accessed 31 May 2015

EGF (2015) From waste to resources http://www.egf.pt/content/index.php?action=detailfo&rec=1885&t=Valorisation. Accessed 29 May 2015

Eurostat (2013) Packaging Waste Statistics. http://ec.europa.eu/eurostat/statistics-explained/index.php/Packaging_waste_statistics. Accessed 20 May 2015

Eurostat (2015) Municipal waste statistics. http://ec.europa.eu/eurostat/statistics-explained/index.php/Municipal_waste_statistics. Accessed 29 June 2015

Fast Lane (2014) GM recognized for driving a global movement for zero waste. https://fastlane.gm.com/2014/03/26/gm-recognized-driving-global-movement-zero-waste/. Accessed 29 June 2015

Finnveden C, Hauschild MZ, Ekvall T, Guinee J, Heijungs R, Hellweg S, Koehler A, Pennington D, Suh S (2009) Recent developments in Life cycle Assessment. J Environ Manage 91:1–21. http://xa.yimg.com/kq/groups/22679040/33961240/name/09+-+Finnveden+et+al.+-+Recent+developments+in+Life+Cycle+Assessment.pdf. Accessed 25 May 2015

Flaten TP, Alfrey AC, Birchall JD, Savory J, Yokel RA (1996) Status and future concerns of clinical and environmental aluminum toxicology. J Toxicol Environ Health 48:527–541 http://www.researchgate.net/profile/Trond_Flaten/publication/14429611_Status_and_future_concerns_of_clinical_and_environmental_aluminum_toxicology/links/00b7d519b96fb1dcb2000000.pdf. Accessed 23 May 2015

General Motors (2011) Chevrlolet Volt spares landfills from Gulf oil spills. http://media.gm.com/media/us/en/chevrolet/news.detail.html/content/Pages/news/us/en/2011/May/0523_volt.html. Accessed 29 June 2015

General motors (2012) GM Scrap vehicle material insulates coats for homeless. http://media.gm.com/media/us/en/gm/news.detail.html/content/Pages/news/us/en/2012/Sep/0910_coats.html. Accessed 29 June 2015

General Motors (2013) GM develops Green Thumb in urban gardening. http://media.gm.com/media/us/en/gm/news.detail.html/content/Pages/news/us/en/2013/Aug/0829-urban-gardens.html. Accessed 29 June 2015

Huang C-C, Ma H-W (2004) A multidimensional environmental evaluation of packaging materials. Science Total Environ 324:161–172. http://ntur.lib.ntu.edu.tw/bitstream/246246/96881/1/10.pdf. Accessed 21 May 2015

Hugger T (2004) Ecotip: Glass – What's the environmental impact? 22 Nov. http://www.treehugger.com/culture/ecotip-glass-whats-the-environmental-impact.html. Accessed 21 May 2015

Ilgin MA, Gupta SM (2010) Environmentally conscious manufacturing and product recovery (ECMPRO) A review of the state of the art. J Environ Manage 91:563–591. http://www.researchgate.net/profile/Mehmet_Ali_ILGIN/publication/38032833_Environmentally_conscious_manufacturing_and_product_recovery_(ECMPRO)_A_review_of_the_state_of_the_art/links/00b4952b7e673537c8000000.pdf. Accessed 24 May 2015

Iyer E, Banerjee B (1993) Anatomy of Green Advertising. Adv Consum Res 20:494–501. http://www.acrwebsite.org/search/view-conference-proceedings.aspx?Id=7499. Accessed 30 May 2015

KAO (2014) Environmentally conscious containers and packaging [EC2, EN1, EN27] http://www.kao.co.jp/corp/sustainability-reports/en/environment/packaging/. Accessed 29 May 2015

Kleindorfer PR, Singhal K, Wassenhove LNV (2005) Sustainable operations management. Prod Oper Manag 14:482–492. http://www.icesi.edu.co/blogs/semillerosi3/files/2013/12/06-04-PK.pdf. Accessed 31 May 2015

Laroche M, Bergeron J, Forleo GB (2001) Targeting consumers who are willing to pay more for environmentally friendly products. J Consum Mark 18:503–520. http://www.diva-portal.org/smash/get/diva2:398010/fulltext01.pdf. Accessed 24 May 2015

Lifshitz I (2014a) Top five packaging trends for 2015. http://www.greenerpackage.com/source_reduction/top_five_packaging_trends_2015. Accessed 30 May 2015

Lifshitz I (2014b) Give the people what they want: four emerging trends in sustainable packaging. http://www.sustainablebrands.com/news_and_views/packaging/ian_lifshitz/give_people_what_they_want_four_emerging_trends_sustainable_pa. Accessed 30 May 2015

Liu G, Muller DB (2012) Addressing sustainability in the aluminum industry: a critical review of life cycle assessments. J Cleaner Prod 35:108–117. http://www.researchgate.net/profile/Daniel_Mueller11/publication/257408730_Addressing_sustainability_in_the_aluminum_industry_a_critical_review_of_life_cycle_assessments/links/00b495282e846e1f94000000.pdf. Accessed 23 May 2015

Maria R, Pujari DD (2010) Mainstreaming green product innovation: why and how companies integrate environmental sustainability. J Bus Ethics 95:471–486. doi:10.1007/s10551-010-0434-0. http://web.a.ebscohost.com/ehost/pdfviewer/pdfviewer?sid=388a1c7d-a049-4384-a6d8-15fe4c10f5ed%40sessionmgr4003&crlhashurl=login.aspx%253fdirect%253dtrue%2526profile%253dehost%2526scope%253dsite%2526authtype%253dcrawler%2526jrnl%253d01674544%2526AN%253d52663915&hid=4112&vid=0. Accessed 30 May 2015

Mehr (2015) Sustainability check for recycling of beverage packaging: reuse better than one-way containers with deposit better than curdside collection schemes (green dot systems) http://www.duh.de/pwc_study_eng.html. Accessed 29 May 2015

Nobel N, Pauu L, Mcminimeel C, Malletti M, Singh J (2009) Packaging trends for bottled water. J Appl Packag Res 3:123–135. http://digitalcommons.calpoly.edu/cgi/viewcontent.cgi?article=1035&context=it_fac. Accessed 31 May 2015

Olson WW, Sutherland JW (1994) Environmentally conscious manufacturing. Japan-USA symposium on flexible automation, conference paper, pp 1035–1042. Japan, 11–18 July. http://www.me.mtu.edu/~jwsuther/Publications/191_PA010.pdf. Accessed 24 May 2015

PETCORE (2014) Europe recycled over 65 million PET bottles in 2013 http://www.petcore.org/news/europe-recycled-over-65-billion-pet-bottles-2013. Accessed 29 May 2015

Powell J (2011) Packaging recycling markets. http://www.epa.gov/osw/conserve/smm/sfmr/webinar4-resourcerecycling.pdf. Accessed 30 May 2015

Price A (2010) GM is using oil soaked booms in Chevy Volt. http://magazine.good.is/articles/gm-is-reusing-oil-soaked-booms-in-the-chevy-volt. Accessed 29 June 2015

PricewaterhouseCoopers PWC (2011) Reuse and recycling systems for selected beverage packaging from a sustainability perspective—aggregation of selected significant findings http://www.bottlebill.org/assets/pdfs/pubs/2011-ReuseAndRecycling.pdf. Accessed 28 May 2015

Reusable Industrial Packaging Association (RIPA) (2011) Climate change benefits of industrial transport packaging reuse. http://www.reusablepackaging.org/sustainability/. Accessed 20 May 2015

Reuse (2014) Re-use more, throw less campaign saves 700 tonnes from the trash http://www.rreuse.org/this-is-a-standard-post-format-with-an-image-gallery/. Accessed 29 May 2015

Richards DJ (1994) Environmentally conscious manufacturing. World Class Des Manuf 1:15–22. http://www.academia.edu/4254363/Environmentally_Conscious_Manufacturing. Accessed 25 May 2015

Sakao T (2007) A QFD-centered design methodology for environmentally conscious product design. Int J Prod Res 45:4143–4162. http://dx.doi.org/10.1080/00207540701450179. http://www.diva-portal.org/smash/get/diva2:398010/fulltext01.pdf. Accessed 25 May 2015

Sarkis J (1998) Evaluating environmentally conscious business practices. Eur J Oper Res 107:159–174. http://www.researchgate.net/profile/Joseph_Sarkis/publication/235791805_Evaluating_environmentally_conscious_business_practices/links/0fcfd512689921fd7c000000.pdf. Accessed 25 May 2015

Tongpool R, Jirajariyavech A, Yuvaniyama C, Mungcharoen T (2010) Analysis of steel production in Thailand: environmental impacts and solutions. Energy 35:4192–4200. http://www.nru.ku.ac.th/KU_NRU_/upload/File/67652c0dcb3a8c4ba99b85f80d2d5495.pdf. Accessed 23 May 2015

Tuckerman R (2005) Packaging the Statistics. 10 Nov. http://www.pca.org.au/uploads/00207.pdf. Accessed 20 May 2015

Wadhwani A (2014) 3 Trends that will define sustainable packaging in 2015. http://www.business2community.com/sustainability/3-trends-will-define-sustainable-packaging-2015-01098694. Accessed 30 May 2015

FSC
www.fsc.org
MIX
Papier | Fördert gute Waldnutzung
FSC® C083411

Kolibri 360 GmbH
Ferdinand-Jühlke-Straße 7
99095 Erfurt, Deutschland
produktsicherheit@kolibri360.de